THE VENUS ECLIPSE OF THE SUN 2012

The
VENUS ECLIPSE
OF THE SUN 2012

A Rare Celestial Event
Going to the Heart of Technology

DAVID TRESEMER, Ph.D.

LINDISFARNE BOOKS | 2011

2011
LINDISFARNE BOOKS
610 Main St. Great Barrington, MA
www.steinerbooks.org

Some of the material in this book was published in *Journal for Star Wisdom,* an excellent forum for many innovative ideas of astrosophy: *astro-* (star) –*Sophia* (the feminine principle of divine wisdom). *Journal for Star Wisdom* is available from www.SteinerBooks.com.

Ongoing research papers, archives of some of the articles from *Journal for Star Wisdom,* "The Oracle of the Solar Cross," discussed herein, and "Tea with Your Mentor," also discussed herein, can be found at www.StarWisdom.org.

Cover illustration and design by Fiona Stewart
Book design by William Jens Jensen

LIBRARY OF CONGRESS CATALOGING-IN-PUBLICATION DATA

Tresemer, David Ward.
 The Venus eclipse of the sun 2012 : a rare celestial event going to the heart of technology / by David Tresemer.
    p. cm.
ISBN 978-1-58420-074-1
 1. Venus (Planet)—Transit—2012—Miscellanea. 2. Technology—
Miscellanea. 3. Astrology. I. Title.
BF1724.2.V45T74 2011
133.5′34--dc23

                                                                          2011041005

eBOOK ISBN 978-1-58420-115-1

CONTENTS

Acknowledgments

My gratitude goes to the members of the StarFire Research Group, who gather in person or on the telephone to discuss the issues within this book, as well as many other issues about how Earth and heavens interact. I am also grateful to Sarah Gallogly for her many editorial suggestions.

NOTE

Knowledge of astronomy and astrology is not necessary to understand the discoveries in this book. For those who know about the system of astrology, the format "12 Leo 8" means 12 degrees and 8 minutes of the sidereal astrological sign of Leo, using signs of equal length. More about the sidereal approach and why it is important for this kind of work can be found in references in the foreword and Appendix C. "J" means Julian calendar, rather than the Gregorian calendar used presently (different countries switching to that way of reckoning at different times, indeed spanning from 1582 to 1926). To repeat, these details in footnotes are meant for the specialist, and need not become obstacles to the appreciation of this book's contents by everyone.

FOREWORD

In our mad rushes around this beautiful planet, we seldom look up to the great wonder of the heavens above us, as we move swiftly through our comfortable realms of space and time. Every so often, you have a clear night without too much city light, and the immensity comes as a shock. What impact does it have? Some people explore that by reading the astronomer popular at the time (Carl Sagan, Stephen Hawking) or visiting a planetarium. Some people consult an astrologer—all that activity up there must have some connection to this planet! Some people—including myself—suggest that you simply find a star and gaze at it, and note the flavor of your feeling response, for comprehension of the heavens is too great for one's thinking. You can only come to grips with this enormity through the heart.

This book has come about because a rare celestial event is occurring, one that will be referenced in years to come. I explain the event itself in chapter 1 (with technical details in Appendix A), then entertain all of its implications and ripples in later chapters. To prepare you for this way of thinking, I would like to orient you to the heavens, to the skies. These great mysteries can be approached through astronomy—the description of the heavens, the naming (*-nomy*) of the different points of light as well as the dark spaces. You can also approach the mystery through astrology—pattern and explanation (*logos*) of the stars (*astra*)—or through astrosophy—wisdom (Sophia, the divine feminine) of the stars (*astra*).

Most people aren't familiar with the term *astrosophy*, so I will most often include it under the term *astrology*. However, astrology has gotten a bad name for itself. In graduate school, I was taught that astrology was superstition. Later I was challenged by peers to test it with my fancy statistics. I did so, and was so impressed that I began to learn the traditions of astrology. While much mainstream astrology seems silly and misleading, I have found a wisdom available about the heavens provided by the discipline of astrosophy, star wisdom, guided by the teaching of *Anthroposophy*—another Greek term, from becoming-human-being (*anthropos*) in wisdom (*Sophia*). This goes back further to the School of Zarathustra, the tradition that informs a sidereal—star-based—view of astrology.

The heavens have several moving parts. The stars themselves, twinkling from their vast distances, measured in the incomprehensible units of light-years away, are called "fixed," though they do move, albeit very slowly. The *planets*, a Greek word meaning "wandering stars," move in regular and ordered patterns in their orbits, coming into and going out of resonant relation with each other. The Sun stands regally as the center of this dance. From the Earth's point of view, the Moon is the most active of all in the heavens, and all the creatures relate to her rhythms. The Earth's rotating through night and day also participates in this grand celebration of life, for isn't life and its enjoyment the point of all of this? A trained astrologer can apply his or her own experience, and the experience of others written down in tomes that go back hundreds of years, to the particular pattern of planets and stars.

Zarathustra divided the heavens into the realms of twelve holy beings, whom he could perceive, and thirty qualities within each realm, making 360 different qualities, from which we get our sense of space (through trigonometry applied to longitude and latitude, as well as to all geometry) and time (through the division of the hours and minutes of the day). We can make maps—which we call charts, as in charting a course through the cosmic seas—of the heavens. One is included in Appendix D for the Venus eclipse event. In that chart you can see Venus (circle with cross beneath) in the sign of the Bull (Taurus, with a circle surmounted by a semicircle, to suggest a bull's head and horns), at 20–21 degrees in the Bull.

Critics challenge astrology as unscientific. "Why can't astrology be more like medicine?" they ask. Appendix C gives a partial response to these criticisms.

I have been interested in the different qualities of each of the 360 degrees. Based on an accurate dating of the experiences of Christ Jesus in the last four years of his life, and the statement by the philosopher Rudolf Steiner that Christ Jesus was in complete accord with all celestial movements in the heavens, I created the Oracle of the Solar Cross. I gathered together all the days that occurred in a particular degree, let's say 5 degrees of the Scales (Libra), then researched the events that occurred on those days in their historical context.[1] Finally I meditated upon those events, dialogues, and scenes and arrived at a word Image that summarized the influence of each degree. I found that the Image amplified by the Sun at that place in the heavens coordinated with the Image directly opposite, right through the Earth, as well as the Images to the east and west horizons, 90 degrees from the vertical.[2]

Typically a Solar Cross reading counsels an individual about the qualities of his or her life, about the challenges and opportunities into which the individual was born. As the Venus eclipse of the Sun is an event that affects everyone, the relevant Images can be plumbed for truths about our time.

---

1.   The astrological placements are based on a sidereal system of reckoning, which is necessary for finding the same stardate for comparisons. In the footnotes, the format 12 Leo 8 means 12 degrees and 8 minutes of the sidereal astrological sign of Leo, using signs of equal length. "J" means Julian calendar.

2.   The resources are recounted in David Tresemer (with Robert Schiappacasse), *Star Wisdom & Rudolf Steiner* (Great Barrington, MA: SteinerBooks, 2007). The events were recorded by Clemens Brentano from visions of Anne Catherine Emmerich (recorded in Anne Catherine Emmerich, *The Life of Jesus Christ and Biblical Revelations*, four volumes, Rockford, Illinois: Tan Books, 1986, reissued 2004) and dated by Robert Powell, *Chronicle of the Living Christ: The Life and Ministry of Jesus Christ: Foundations of Cosmic Christianity* (Hudson, NY: Anthroposophic Press, 1996).

For more information on basic astrology, I recommend my own *Star Wisdom & Rudolf Steiner* (with Robert Schiappacasse). For a sense of which zodiac to use—sidereal or tropical—I recommend *The Astrological Revolution* by Robert Powell and Kevin Dann, and for more extensive background, *The History of the Zodiac*.[3]

Finally, for those who wish to understand where this work can lead, I offer a Star Wisdom Credo.

---

3.   *Star Wisdom & Rudolf Steiner*, ibid.; Robert Powell and Kevin Dann, *The Astrological Revolution* (New York: Lindisfarne, 2010); and Robert Powell, *History of the Zodiac* (San Rafael, CA: Sophia Academic Press, 2006).

Star Wisdom Credo

SPACE has two modes: distance and resonance,
the first familiar, measured in inches or meters,
or the number of steps from the base of the mountain to its top,
the second also familiar, connecting one thing with another thing,
by common vibration, by the feel of sameness or harmony.
Beyond distance, dull and plodding, is frigid darkness, then nothingness.
Beyond resonance, tone-filled and flying,
    is warming brilliant light, then unity.
The "I"-consciousness of human beings must tread a middle path,
between the extremes of nothingness and unity.
Though science cleaves to distance, the psyche knows resonance.
The living beings of the stars are right here with us through resonance,
and can help us find our path through space.
TIME has two streams:
From past to present and beyond to the future
flow crusted pictures of history that seal karma's working.
From future to present and beyond to the past
flow the picture-less urges of destiny.
The streams collide, creating dangerous eddies and gaps.
In the middle live the healthy rhythms of life process:
heartbeat, breath, waking, sleep, digestion,
seven-year life stages, relationships,
and contraction to death.
The cycles of the Sun, Moon, and wandering planets
    govern this middle realm.
The sparkling stars invite us to stretch past the streams of flow,
toward immediacy and toward eternity.
Only in the great mansion of the active HEART
can we find the strength
to know SPACE and TIME as mentors of freedom
rather than as chains of enslavement.

1

THE ECLIPSE OF THE SUN BY VENUS

Much has been written about the end of the great calendar of the Mayan peoples, signaling the end of time and perhaps the end of life as we know it, even giving it an exact date of December 21, 2012, winter solstice in the northern hemisphere, when the Earth and the Sun are supposed to align uniquely with the center of our galaxy. This alignment (Sun—Earth—[near to] Galactic Center or Earth—Sun—[near to] Galactic Center) is, however, not rare. The Earth and Sun have aligned with a place near (never on) the Galactic Center twice a year since time immemorial. They make that alignment on the dates of winter and summer solstice twice a year every year for over thirty years, after which the dates for this alignment move on past December 21 and June 21. Appendix A explains the celestial dynamics a bit further.

The most important rare astronomical event of 2012 is the movement of Venus before the face of the Sun, which occurs in pairs eight years apart with over a hundred years in between the pairs. Before 2004, the previous eclipses were in 1874 and 1882. The next one will be in 2117. The event in 2012 will occur on June 5–6. (More details are given in Appendix A.)

Here's how the lineup appears on that day:

20–21 degrees of the Bull (Taurus)
—Sun—Venus—Earth—
20–21 degrees of Scorpio

This is truly rare, and we can ask what impact we might possibly feel from it. To answer this question, we have to build up our understanding of this eclipse by looking at four factors: 1) the Sun; 2) Venus; 3) their relation through eclipse; and 4) where this occurs in the heavens, 20–21 degrees of the Bull, further out from the planets, in the periphery of what we can observe.

Does this lineup accentuate the effects from the starry periphery by combining influences, or does the lineup diminish those effects by placing the body of Venus in the way of the Sun? Based on consultations with clients, my approach has always included both, understanding such alignments as increased turbulence of the themes living in the heavens and focused through the Sun. The individual characters of Sun and Venus are both enhanced and challenged by their alignment.

The Sun

The Sun is the most important feature of our solar system, its greatest power of warmth and light. We always note the presence of the Sun, and call it daytime. We follow very closely and behave quite differently in the different parts of the cycle of the Sun: day, night, dawn, and twilight. The Sun acts as a focalizer, a laser beam to the Earth that amplifies what lives behind it in the starry realms. The Sun moves, each day giving us amplifications of a different slice of the heavens.

The Sun focuses in both directions. Not only does the Sun amplify to the Earth what lies on its other side in the starry heavens. The Sun also takes what the Earth has to offer and amplifies its import into the heavens beyond the Sun. This adds to what the stars already hold as world memory in that zone of the zodiac.

When you are born, the Sun imprints you with the qualities living in the heavens, including the world memories in that particular degree. Among those memories is the memory of your birth. That's why you celebrate your birthday at the time when the Sun returns to the same place in the heavens a year later: Your memory of original intention for your life is jogged on that day. "Happy Birthday!" means "May you feel in your bones the initial impulse that brought your lovely and unique being into this world!" From the individual's point of view, there is one special day of the year when this can be felt more than usual.[1] You can experience the Sun as holding something truly special for your most intimate and inner self, for your soul—holding a memory of intention for this lifetime, and cheering you on in the fulfillment of these intentions.

From the point of view of the Sun, every day is a special day, with unique qualities that it focuses to the Earth. We can participate in the birth day celebrations of different qualities for the Earth each and every day.

Venus

The second planet from the Sun is a messenger of the Sun. It has particular qualities when in relation to other planets and to particular places in the heavens.

The physical cycles of Venus are described in Appendix A.

What do we know about her feeling qualities and her influences? What is it like to be in her presence? How do we feel her power when she is stronger? We study this intimately in chapter 7. Briefly, Venus can be understood as influences to see and listen deeply—Venus listens. Venus stimulates the desire to explore relationship, to find one's soul groups or karmic affiliations. Venus focuses beauty, and attention to relationships—the many tones of beauty in a relationship with another human being, and groups of like-minded human beings. Venus is grace; she is sex; she is attraction;

---

1. At http://www.StarWisdom.org, we give specific recommendations for a birthday party at noon, just between you and the Sun, preceding any social party that you might hold on that day.

she enlivens Sophia, feminine divine wisdom.

Botticelli's *Birth of Venus* expresses grace in movement, here rendered dark because of the power of the Sun behind her.

She does not stand straight and symmetrically. Botticelli's rendition suggests undulating movements. Nakedness and modesty are expressed at the same time. We participate in a celebration of the human body, its beauty and capacity. Imagine now that Venus inspires you to find the groups of other human beings who pursue specific aims in service of humanity, fired by the power of sexual passion, yet devoted to the expression of beauty.

Relationship—"Aspects"

From the Earth's point of view, Venus can have several relationships with the Sun, called aspects. From the Earth's point of view, as Venus lies so close to the Sun, it can't be opposed to the Sun, or even square (at right angles) to the Sun. It can go as far as semi-square (45 degrees) but no further. From the Sun's point of view, Venus can indeed extend to a square (right angle, 90 degrees) to the Earth. In astrosophy, we view the square as the relationship of Challenge, opposition as Complement, and conjunction as Gift. In this study, we are interested in the times that Venus lies conjunct to the Sun, in the relation of Gift as enhanced by the great Sun. As all the planets tend to travel a bit above or below the Sun's path, conjunction often means standing a bit above or below. We imagine the picture of standing next to, even a large body and a smaller body standing next to each other, as gifting each other, and combining influences to gift to the Earth. An eclipse, however, is a very special case of a conjunction.

ECLIPSE

Sometimes a heavenly body does not pass above or below the Sun on their shared path through the heavens, but can come directly between the Sun and the Earth. We are most familiar with eclipses by the Moon, though Venus and Mercury can do this also. No other planet or heavenly object ever stands between the Sun and the Earth.[2]

The Moon eclipses the Sun about twice a year. Traditionally, this was thought to imperil the Earth: "Our masters taught: When the Sun is in eclipse, it is a bad omen for the entire world, for all of it."[3] Does this also hold for a Venus eclipse? Let's think of it this way. From the point of view of the individual human being, the Earth eclipses the Sun for half of each day—we call that eclipse night. Ponder the difference between day and night in your behavior and outlook. Ponder the crazy things that you sometimes do at night, the basis for the Russian proverb "Mornings are wiser than evenings." Your relationship to your life's intention becomes clearer during the day. Just for the sake of safety (yours and others'!), you lay your body down at night into sleep, and take the time when out of the Sun's protection to journey into worlds of dreaming, while the body is slowly rebuilt and regenerated.

---

2. An exception in fiction is *The Black Cloud* by Fred Hoyle (New York: Harper, 1957), where a massive cloud first comes behind the Earth, thus increasing the retention of heat, and everything burns up. Then it passes between the Earth and the Sun, and everything freezes. This is an exception that proves the rule—seldom does anything intercede between Sun and Earth.
3. Hayim Nahman Bialik and Yehoshua Hana Ravnitzky, *The Book of Legends: Legends from the Talmud and Midrash* (New York: Schocken, 1992), p. 764, no. 43.

Night—the eclipse of the Sun by Earth—can bring crazy behavior. Yet night can also bring creativity and new ideas, sometimes notions that, in the light of day, evaporate, and you wonder what you thought was so special. And some of those creative impulses remain, and inform the rest of your waking life. When you really understand the powers of the night versus the powers of the day, you can begin to comprehend Venus coming before the face of the Sun in relation to Venus' operation on your consciousness.

Venus orbits the Sun every 225 days, just under eight months. From the Earth's point of view, watching the Sun and Venus progress on their usual path through the course of time, Venus usually crosses the Sun's position on the path above or below the Sun's bright disc. In 1874 and 1882, Venus went before the Sun's face, and again in 2004. The second of the pair will occur on June 5–6, 2012. The Venus eclipse of the Sun will not recur until December 2117 and 2125, in another pair eight years apart.[4]

Eclipses of the Sun by the Moon, where the Moon blocks the light of the Sun for a short time over a small part of the world, are known for their effects after they occur. Astrologers over the ages have found that these effects also precede the event itself. Much as one would anticipate a finish line when completing a long race, especially when the finish line comes into one's view, the effects of a conjunction (or other aspect) seem to occur before an event, as well as after. The themes from the eclipse of the Sun by Venus will affect all of 2012 as well as some years afterward—and perhaps before as well.

Technically termed a "transit" or an "occultation," Venus' crossing before the face of the Sun appears to us as a ring-shaped, or annular, eclipse. When the Moon passes exactly in front of the Sun, we experience an eclipse, the Moon often exactly matching the size of the Sun. When the Moon in its

---

4. The significance of the eight years is shown in the figure in Appendix A, in which Venus creates a beautiful five-pointed star from the perspective of Earth. The next eclipses of the Sun by Venus will occur Dec. 11, 2117, and again Dec. 8, 2125.

orbit is further away from the Earth (and thus appears smaller to us), it does not cover the entire face of the Sun. Then we see the eclipse as a dark core surrounded by a ring of fire. In terms of effect, the word "eclipse" captures the coming event most accurately.[5] Venus before the Sun will appear as a small presence against an immense fiery background. However, like a pebble in your shoe or a love letter in your pocket, the effect is disproportionate to the relative sizes of the things.

From the teachings of Anthroposophy,[6] a planetary sphere—the form delimited by the path of its orbit and extending up and down from that great circle—has a certain quality of energy intrinsic to it. The Venus sphere includes relationships to groups, to the group to which you belong in your efforts to accomplish something of significance in this life, those who are your brothers and sisters in spirit, and who have similar ways of working in the world.

To Anthroposophy, a planet is seen as the densest representative of its sphere, the least potent place of the sphere, in a sense blocking the energy of that sphere.[7] However, consider a smooth flowing stream. A large rock is put in the middle, whose diameter is a hundredth of the width of the stream (the same proportion as Venus to the Sun). The river continues to flow, but there are now eddies and whorls where it had been smooth before.

---

5. The word *transit* is used in astrology to mean entry of the moving planets into one of many different kinds of relation to positions of celestial objects at your birth, including aspects of conjunction, opposition, square, and many others. Thus it is much broader than the astronomical use of the term to mean "lying in front of." The word *occultation*, though correct, has that pesky word "occult" in it, which has gotten a bad reputation.

6. *Anthropos*, possible human being, *-sophy*, divine feminine wisdom. Anthroposophia guides the Anthroposophical Society based on the teachings of the philosopher and spiritual teacher Rudolf Steiner (1861–1925).

7. Rudolf Steiner, *Spiritual Guidance of the Individual and Humanity* (Hudson, NY: Anthroposophic Press, 1991), lecture 3, June 8, 1911, p. 61: "If in a particular horoscope Mars is over Aries, this means that certain Aries forces cannot pass through Mars but are weakened instead."

The delivery through the Sun of the star resonance that lies behind it (as well as the world memories stored there) has been altered. The quality of life downstream takes on a different quality. This is a picture of turbulence—the energy streams moving more swiftly in some places, and more slowly in others. Thus we see both accentuation and diminution of the influences of the zodiacal periphery, the Sun, and Venus.

Thus the eclipse will stimulate in turbulence the Venus question "To what group does my 'I'—my individuality, my sense of self—belong? What is the purpose or work of my group of human beings? How can I participate actively and further this purpose?" Will the Venus eclipse make these questions more anxious? Or will the influences of Sun and Venus combine to answer these questions? Will Venus amplify the Sun's delivery of what lies behind it? Or will we have a diminution of the Sun's message?

Venus before the Sun recalls the image of the woman "clothed with the Sun" from chapter 12 of the *Book of Revelation*. The Sun accentuates the powers of Venus, pictured on this page in a version of the Virgin of Guadalupe.

LOCATION OF THE ECLIPSE

What is the location—the heavenly address—of this eclipse? What, empowered by this alignment, streams from the periphery of the zodiac?

The Sun with Venus before it will stand between 20 and 21 degrees of Taurus—in the system of astronomy, which agrees with that of astrosophy.[8] What is the nature of that location? The Oracle of the Solar Cross gives thematic word images for each of the 360 degrees of the zodiac, as in a cosmic dream.[9] We normally give a Solar Cross reading in relation to an individual's birth, when he or she breathed the first breath. Unlike a personal birth, the Venus eclipse of the Sun will be experienced by everyone on the planet. We will explore the themes that will be stimulated by this celestial event in the rest of this book. We will also look at personages who were born with the

---

8. The "tropical" system in use in the West sees the heavenly addresses differently. Our approach is sidereal, that is, star based, *sidus* meaning star, in contrast to the oft-used tropical system, which is a seasonal calendar based on the important turning points of the seasons (March 21, Dec. 21, etc.). These two systems for mapping the heavens are contrasted in *Star Wisdom & Rudolf Steiner*, which recommends that the tropical system, which was astronomically correct two thousand years ago and has slowly strayed from that accuracy, use other names for the months based on the seasons, such as "month when blueberries sweeten." The sidereal system that I use takes its foundation from very ancient sources, described in Robert Powell, *History of the Zodiac*, and *Hermetic Astrology* (3 vols., San Rafael, CA: Sophia Foundation Press, 2006). I emphasize here the location in relation to the stars, thus five degrees past the central star of the constellation of the Bull (Taurus), namely Aldebaran, the Eye of the Bull, at 15 degrees and 3 minutes of the Bull. This in itself is a critical placement, as Aldebaran, and its exact opposite, Antares, the Heart of the Scorpion, at 15 degrees and 1 minute of the Scorpion or Scorpio, form the traditional basis for division of the heavens—see Robert Powell, *History of the Zodiac* (San Rafael, CA: Sophia Academic Press, 2006).

9. The Oracle of the Solar Cross is introduced in David Tresemer, with Robert Schiappacasse, in *Star Wisdom & Rudolf Steiner: A Life Seen Through the Oracle of the Solar Cross* (Great Barrington and New York: SteinerBooks, 2007). As explained more fully in *Star Wisdom & Rudolf Steiner*, based on research of the daily events in the life of Christ Jesus, I created word images of the qualities of each of the 360 degrees of the zodiac.

Sun in that degree (chapter 5), as well as world events and deaths that might condition our ability to work with the qualitative themes active there (chapter 6). In a regular Solar Cross reading, the Image for this degree (given in the next chapter) is read and pondered for hints about one's life. Because of the rare event of the Venus eclipse in 2012, everyone on the planet will be related to the content of this Image in the Bull (and its exact opposite in the Scorpion, revealed and pondered in chapter 3).

People often seek guidance from celestial movements, consulting astrologers to mediate that guidance. Most often they ask for practical advice, relating to the problems that they face in their lives. Magazines offer "six steps to better digestion" or "eight steps to greater intimacy." Another kind of knowing resides in the mythic, symbolic, or poetic: the realm touched by the Images in the Oracle of the Solar Cross. In a healthy life, the two realms circulate and inform each other. The mythic/poetic pictures of a Solar Cross image are more akin to dreams, not imparting immediate practical advice but inspiring with a feeling tone or quality. Each of us has dreams that occur within our personal sphere. An Image for a particular location in the heavens is a cosmic dream in a world sphere that includes my sense of "I"—I live within the dream. Eventually "I" will find the interaction where "I" lives within the dream and it within my "I". On to the dream of 20–21 degrees of the Bull.

THE GREAT KING
AND HIS MAGICAL IMPLEMENTS

We begin at the periphery, what lives furthest from the Sun in the zodiac. The Sun points us to one particular degree, 20–21 of The Bull (Taurus), which we can consider a gate through which come qualities and themes. Each of the gates in the Oracle of the Solar Cross has a long word Image and an abbreviated word Image, then commentaries on the Image. Here is the longer Image that streams through the Sun at 20–21 degrees of The Bull, Taurus:

> Seven philosophers recall with pride the great king of their lineage, Jamshid, who had gifts from heaven: a golden dagger, a flying firebird, a cup of seven circles, and a golden ring with an emblem of power. With the golden dagger, the king measured the land, separating parcel from parcel, and plowed the soil. From the flying firebird he directed fire for all humankind, intense fire to smelt metals from rock, from which could be made many useful things, as well as the transformative fires of cooking. The waters of the cauldron thwarted disease, old age, and death; in the waters he could see whatever was happening anywhere and anytime. The golden ring could imprint the command of divinity, even expanding and renewing the Earth itself. The king prolonged the life of all people and animals in his kingdom for hundreds of years. A wise visitor reminds the seven philosophers that there is a greater king, whose tools are kept in the heart. Does he mean Zoroaster, does he mean Melchizedek, does he mean Inanna or Solomon? Or is there perhaps a greater king?

The Image develops in parts. Though the end of the Image extols the ability to use fewer divine implements, the beginning is full of them. We need to understand these implements clearly so that we can later internalize their functions in our hearts.

THE SEVEN PHILOSOPHERS

This discussion takes place among seven philosophers, an echo of the Seven Rishis, or the Seven Sages, who in Hindu tradition have overseen human development since its inception.[1] We can perhaps feel them in the background overseeing this discussion.

JAMSHID THE GREAT

The seven discuss Jamshid. Zarathustra or Zoroaster—the author of the ancient text named the Zend Avesta, now the scripture of the Zoroastrian religion, and also the earliest incarnation of the series of Zarathustra teachers who have led the school of astrosophy—described in that ancient book how he put a question to the prime creator of all, Ahura Mazda. Zarathustra asked if Ahura Mazda had spoken to another human being before Zarathustra. The creator answered, "Jamshid." Ahura Mazda had given Jamshid directions to care for the world. Jamshid replied to Ahura Mazda, "Yes! I will make thy worlds thrive, I will make thy worlds increase. Yes! I will nourish, and rule, and watch over thy world. There shall be, while I am king, neither cold wind nor hot wind, neither disease nor death."[2] Ahura Mazda awarded Jamshid with tools to assist him in this task, those that are enumerated in the Image at

---

1.   The Seven Rishis are spoken of in Hindu texts, and in pre-Hindu Vedic texts, and in the Trans-Himalayan wisdom of Alice Bailey theosophy. They are often identified with the seven stars of the Big Dipper or Great Bear. The seven walking in discussion can thus be seen as an earthly counterpart of the great seven in the heavens.

2.   Fargard II of the Zend Avesta, First Part, chapter 5 (14).

20–21 of the Bull, Taurus. (See Appendix B for the many names of Jamshid as well as other details about this historical figure.)

GOLDEN DAGGER

The first tool that Jamshid received from heaven was a dagger. It had various functions, in the elements of air and earth.

The airy function of this special dagger—golden, indicating its origin in the Sun—was its ability to measure the land. Most governments have a bar of metal kept at a cool temperature with two tiny scratches on it to show the definition of the important measures of the land, as in the twelve-inch foot or the meter. Jamshid's knife included the royal measures and much more. It showed the proportional interrelations of numbers, including the trigonometric functions necessary to determine, through triangulation, the survey of the land. When the Nile flooded each year, all boundaries between the fertile fields of Egypt were lost. In the spring the priests took the royal measure from the king and applied it to the re-measurement of the fields. The ruler of the land held a ruler. The rod, staff, and scepter of kings always held a measuring function, defining the "sacred foot" or "sacred cubit" or "sacred yard," plus other measures that helped in the trigonometry necessary for surveying the land.[3]

The knife functioned in the element of earth to cut along the boundary lines between plots of land, then separate one side from another. Cutting separates.

---

3. Robert Lawlor goes into this in detail in *Sacred Geometry*, based in part on his translation (with Deborah Lawlor) of Schwaller de Lubicz's great masterpiece, *The Temple of Man* (Rochester, VT: Inner Traditions: 1998), which emphasizes the foundations for sacred geometry in volume 1. The film *Prince of Persia* imagined a golden dagger that affected time. We will see the play with time in Jamshid's seven-ringed cauldron. An annual anthroposophic journal, *The Golden Blade*, affiliates itself with Jamshid's tool and names these themes in its purpose.

Thus the tool of discernment is often pictured as a knife or sword.[4] It can also be used to stir the soil, to break up the hard earth and dislodge weeds so that crops can be grown. For millennia people had used digging sticks, with fire-hardened points, to pry up roots. The brilliance of the plow was to attach a pointed stick to animal power so that it probed the soil. Discussions about the right way to stir the earth took people's attention for thousands of years. The first recorded theatrical drama from millennia ago has an actor as Hoe challenging the ascendancy of an actor as Plow. Hoe and Plow enumerate their own good qualities and the detriments of the other.[5] In this same tradition, Thomas Jefferson wrote treatises on the best plow design. So you see the topic of how to cultivate the earth wasn't just up to agricultural engineers as it is in our time. These questions belonged to everyone and were hotly debated. The best kind of plowing involves stirring rather than turning the soil upside down. At an agricultural museum in Denmark, I once guided a plowing implement with a piece of wood, hardened in the fire, four times the length of my thumb, attached to a framework pulled by two oxen. Back and forth we went, stirring up the field, separating weeds from the earth to make place for plants that we had chosen. One can imagine the golden dagger, indestructible because it comes from divinity, attached to such a framework to cultivate the soil.[6]

Not until groups of humans could create a surplus of food, in the form of stored grain, could they overcome the daily struggle and create civilization as we know it, with arts and sciences and theatre![7]

---

4. As in Rev. 1:16: "From his mouth came a sharp, two-edged sword," by which one understands an ability to cut away truth from falsehood, the true meaning of criticism (rather than cut the messenger into pieces).
5. The script of this oldest of plays can be found at http://www.hort.purdue.edu /newcrop/history/lecture09/lec09.html.
6. I have also guided a tractor with a moldboard plow attached. The function is the same. The definition of "acre" is the amount of land that one can cultivate in a day with a moldboard plow pulled by two oxen. This is another way in which the golden dagger as plow relates to the definition of space.
7. Axel Steensberg has made this argument in various works.

I worked with a group in Melbourne to find physical gestures that summarize the capabilities of the golden dagger to measure near and far, in short, to define Space; then to cut, to separate, to plow the earth, to manipulate the stuff of Space. Indeed, the golden dagger indicates every extension of the hand, in a word, all technology. Thus the primal gesture is that depicted by Stanley Kubrick in *2001: A Space Odyssey*, the ape who learns to pick up a jawbone of a skeleton, and turns it into the first tool...which is a weapon...which then becomes a spaceship approaching the Moon. Finding gestures helps to understand this divine gift and its use by human beings.

FIREBIRD

Divinity gifted a firebird. Fire can then be used for creating things. One begins with the low temperatures, with cooking, including baking and drying. This includes the ignition of wood as in a campfire, or gas in a kitchen stove, or the microwave oven which excites molecules of the food so much that they heat up.

We can increase the temperature and make great volumes of steam, enough to fashion wood and small trees to make them pliable for furniture. We can boil water to make soap and tea and dyes. As the temperature increases, it makes possible the working of copper, the blowing of glass, the smelting of the higher-temperature metals such as iron, and the making of steel. I recall visiting a steel factory and seeing the molten steel poured into immense molds. My guide explained the sparks splashing nearby, "Watch out! Those sparks are not like sparks from a campfire. Those sparks will go right through your pants and your leg, and won't stop until they hit your bone." All the metals that we use come from this application of intense fire.

Light is made by electricity through a medium free of oxygen. The firebird directs the fires of electricity and light.

Mastering the element of fire in its power over earth, and the power of light, the king can direct the firebird wherever he chooses.

Rudolf Steiner observed that birds focus the warmth of the world—the warmth ether—and they focus the process of thought: "The bird is the flying thought."[8] Focusing attention on something warms the thing. Intense focus acts as a laser and can, in those trained to do so, cause spontaneous combustion. Focused thought is related to the firebird.

In the group work in Melbourne, we summarized the firebird with the physical gesture of flicking fire to a specific place, as in a focused abracadabra, the laser-like powers of focused attention.

MAGIC CAULDRON

The cauldron had two functions: relieving the people of disease, old age, and death, as its waters had powers of immortality; and permitting the one gazing into the waters to see far and wide, anyplace and anytime. These functions connect with one another. By banishing distance and time in what one experiences with one's senses, one is freed from their constraint. One has a taste of immortality.[9] This was the first television! It permitted one to see far and wide beyond one's immediate surroundings.

One works here with the elements of earth in the substance of the container and the element of water within the container. The king could look into the water of the magic cauldron and see whatever was happening wherever he chose. The water of the cauldron has a reflective quality, as well as a quality of depth. Gazing into water frees you from the immediate stimuli

---

8. Rudolf Steiner, *Harmony of the Creative Word: The Human Being and the Elemental, Animal, Plant and Mineral Kingdoms* (London: Rudolf Steiner Press, 2001), pp. 6, 83. The most fiery bird is the eagle.
9. The four treasures of the Tuatha De Danaan included the Cauldron of Dagda, which had the same powers. Note that Dagda comes from *dhagho-deiwos*, shining divinity, thus the same meaning as Jamshid.

of the senses and permits you to think about something that is not in front of you, that is, to reflect on something you saw at another time and place, or to imagine something that you haven't experienced before, or to become a philosopher, thinking about abstract concepts and what lies beyond concepts. Does one look at the pictures playing across the surface of the water? Or does one peer down into its depths, as did Nostradamus, to see pictures from other times and places?[10]

Water in a large vessel slowly turns because of the Coriolis effect, a result of the rotations of the Earth. This turning creates a weak electromagnetic field. Though weak, this energy can cumulatively over time affect growth processes nearby. The great psychologist Mesmer used a large cauldron of water in his house, as did Cagliostro, whom we will meet later. The potions in the water can thus be broadcast to a moderately large area in the vicinity. The great cauldrons of old, such as the very large one built at the Temple of Solomon, may have had such a purpose—broadcast of specific qualities of energy to the city surrounding the temple.

"Seven-ringed" may refer to the seven planetary spheres.[11] The cauldron may be as large as the whole heavens. It is creation itself, into which we gaze and are enchanted.

One of the uses of the cauldron is baptism, which extends one's life by awakening the connection to spiritual worlds. Water made holy through repeated blessing becomes the medium through which a kind of immortality

---

10. Rudolf Steiner described Nostradamus in his room with the cauldron of water in "Prophecy," *Astronomy and Astrology* (Forest Row: Rudolf Steiner Press, 2009), 15–42, lecture of November 9, 1911, and cited by William Bento, "Prophecy: A Cauldron of Controversy," *Journal for Star Wisdom 2011*, pp. 41–48.

11. Giorgio De Santillana & Hertha von Dechend make this claim about Jamshid's cup, in *Hamlet's Mill: An Essay Investigating the Origins of Human Knowledge and Its Transmission Through Myth* (Boston: David Godine, 1977), p. 47. The frontispiece there gives an excellent illustration of a seven-ringed cup, made of the spheres of the heavens.

is given to the aspirant, either water poured over the head or full immersion in the water. Though trivialized in many places, when taken seriously, this becomes a profound ceremony for the baptized one as well as all the witnesses.

What is the gesture of function and use for the cauldron? In our Melbourne group, we gathered the arms into a circle, a large horizontal "O" shape, forming the rim of a cauldron, then stared deeply down past the rim, as into water, deep into the hidden realities of distant times and places, where we could imagine the extension of eyes and ears through space and time.

DIVINE EMBLEM

Jamshid gave freely of the magic cauldron, so that no people or animals died. There was no ostensible disease or death in people and animals. But the world eventually became overpopulated and grossly polluted. The plant and mineral kingdoms were sorely stressed, and the psychological diseases of overcrowding must have been rife. The Image exactly opposite to this one, at 20–21 Scorpio (which we shall see in chapter 3), is full of disease, despair, and death. Anyone born in this particular degree must struggle with this polarity.

After three hundred years, the prime creator Ahura Mazda warned Jamshid that this couldn't go on, and urged the king that he send a good portion of animals and people to the realm of death, for the imbalance had to be rectified. Instead Jamshid set the divine emblem on the golden ring to the Earth and commanded that the world increase in size. The Earth stretched out into a larger globe. The king worked actively with the stuff of Earth itself, interesting especially as this Image lies in the earth sign of Taurus.

Jamshid ruled another six hundred years, until the Earth again was overrun and Ahura Mazda warned him that Ahriman, the lord of the deep and the dead, needed people and animals in his realm, whose passage there had been denied by the defeat of death. Jamshid again set his ring with its magic emblem to the ground, and the Earth expanded. He ruled another nine hundred years, and had to expand the Earth once more, but then he became too identified with these works as his own works. He was no longer serving the creator. He lost the favor of Ahura Mazda, and died.

Before he died, sensing a disaster was coming, not only his own death, but the visitation of disease and death on humans and animals, Jamshid did what others have done since his time. He built a great underground facility, two miles long and two miles wide, on many levels, wherein he hoped to house many people. Perhaps they are still there.[12]

The notion of a slowly expanding Earth, rather than a sudden agglomeration that has remained stable, has recently been presented with much compelling evidence.[13] The story of Jamshid may well encode the facts of Earth's evolution, including the diminishing rate of change, at first three hundred years, then six hundred, then nine hundred, showing that the Earth has slowed down in its rate of change. The notion that a powerful king tries to respond to inadequate resources for a growing population— read *over*population or exceeding the carrying capacity of the Earth—make this a very modern theme.

What is the gesture that we discovered for the operation of the golden ring? You press the symbol into the stuff of creation, and the stuff responds. In this way, it parallels the mature Siddhartha's, that is, the

---

12. The tale says that Jamshid created the large underground area for two thousand people. Rudolf Steiner has a helpful view of numbers like this, as in "Feeding of the Five Thousand," as follows: The thousand means a large number, and the modifier, in this case, two, tells us about the quality of energy and the age to which this is related, in this case the second age, the Persian Age, that of Zarathustra. We are presently in the fifth age in that system. Rudolf Steiner, *According to Matthew* (Great Barrington, MA: Anthroposophic Press, 2003), lecture 10. Thus, "two thousand" means a large number of people in the time of the great Persian civilization more than six thousand years ago.

13. James Maxlow, *Terra non firma Earth* (oneoffpublishing.com: Oneoff, 2008), and various Internet presentations on his well-researched thesis.

Buddha's, "earth-touching gesture," wherein he pressed a finger to the earth as an act of will to exist, after which the demons left him alone. The modern counterpart of projecting one's intention outward would include all symbols, logos, brands, as well as electronics, through which people project out their personal intention many times a day. Projecting out one or two hands with intent to alter thought or substance summarizes Jamshid's powers through the ring.

We marvel at the wonderful tools of the great king Jamshid, used for the good of humankind. And yet a question arises: Are these tools always used for the good?

3

TECHNOLOGY AND THE HEART

In the Image at 20–21 degrees of the Bull, we celebrate technology and its origins! Here are wonderful capacities that we now have, given as gifts by divinity. With gratitude, we celebrate our greater ease in living because of these tools.

What does the great king Jamshid do with that technology? He serves humanity. He eases the difficulties of life, increases our capacities to make things that ensure our comfort, and extends our life in this more comfortable world.

THE SHADOW SIDE OF TECHNOLOGY

Yet there is a shadow side to this better world. As Jeremy Rifkin once remarked, the single word that thrusts humanity to its downfall is "convenience." For greater ease, some things are enhanced, while other things are sacrificed, wasted, or ignored. Who can argue with Jamshid's good intentions when he banished disease, old age, and death? Isn't that the thrust of modern medicine? However, Jamshid created the same prison that Siddhartha's parents created for their son. As king and queen, they banished from the central city of their kingdom any evidence of toil, disease, old age, and death. All appeared happy for their darling child. Likewise, the Western world exports its poverty to distant places that can't be observed so that wealthy people can have cheap shoes, cheap T-shirts, cheap everything. Siddhartha had to burst out of the bubble engineered by his parents to mature as a human being, to tread the path to becoming the Buddha. He had to interact with all the phenomena of life.[1]

Modern medicine promises youthfulness through cosmetic surgery and increased longevity through organ replacement.[2] After some thousands of years, Jamshid's story has visited human society again. Is this a good development?

Technology is amoral. The same tool can be used in many different ways. Take a knife, for instance. As the folksinger Lead Belly said, "You take a knife, you use it to cut the bread, so you'll have strength to work; you use it to shave, so you'll look nice for your lover; on discovering her with another, you use it to cut out her lying heart."[3] With technology, we have to wrestle with the issue of morality.

Zarathustra named a naturally occurring "perfecting principle" that works unceasingly on the development of humankind toward the good.[4] Do tools serve the perfecting principle or detract from it? A tool in itself is inert and neutral, but can be used in many ways. Therein lies its danger. It depends on its master to guide its use, for good or ill.

1. Rudolf Steiner saw the same pattern with the Essenes, who banished all temptation and sin to outside their community. They did not transform the difficulties, merely transferred them to others, thus making life outside their community more difficult. Rudolf Steiner, *The Fifth Gospel* (London: Rudolf Steiner Press, 1985).
2. The technology observer Patrick Cox, in a lecture to the Agora Symposium 2009, has said that advances in stem-cell replacement of organs, especially in the research centers in Russia and Korea, where other scientists' patents are ignored, has brought us "functional immortality." The television series *Torchwood* touches upon the dysfunction of widespread immortality in the episode "Miracle Day."
3. Cited in David Mamet, *Three Uses of the Knife* (New York: Vintage, 2000), p. 66. Lead Belly spent much of his life in prison for murder and attempted murder, only rescued by the musicologist John Lomax, whom Lead Belly also tried to kill with a knife. The Sun at Lead Belly's death lay at 20 degrees and 13 minutes of Scorpio, exactly opposed (and thus complementary) to Jamshid's knife.
4. Rudolf Steiner emphasized the perfecting principle in his lecture on Zarathustra, in *Turning Points in Spiritual History* (Great Barrington, MA: SteinerBooks, 2007, original lecture January 19, 1911), p. 14.

We have to ask if technology assists or detracts from the perfecting principle. Most people feel that we don't have a choice. Human beings will exercise their ingenuity and create new and more powerful tools—including bigger and more sinister weapons; there's no stopping it or directing it. This is an admission of powerlessness, that is, ineffectiveness of will. Yet look at the oldest society on Earth, that of the aborigines of Australia. From communion with spiritual beings, what can be termed the Dreamtime or Dreaming, aborigines learned that complex tools were inappropriate for the sanity of their personal and collective lives. In their wanderings through a living landscape, they carried only a few essential tools. The perfecting principle was practiced in complicated ceremony, and through the expression of beauty.[5]

TECHNOLOGY AND THE WILL

We can divide human experience into three realms of thinking, feeling, and willing, in the provinces of head, heart, and hand. Normally we are aware of our thinking—concepts built of perceptions—somewhat aware of our feelings, and little aware of our willing, our deeds and actions. This deserves contemplation. When I realized the truth of this imbalance, I was stunned, and continue my amazement. I know what I think. I define myself by what I think. If I can be aware of feeling, I experience a kind of victory of expansion, for all of beauty then opens to me. But to be aware of my willing, what I am doing, how I am acting, is rare. Often I experience that my mind observes my actions, usually after the act.

An excellent illustration can be found in baseball. Studies have shown that the time between the pitch of the ball, modified by its change of course as in a curveball, and its passage over home plate is too short to allow the

---

5. Robert Lawlor, *Voices of the First Day: Awakening in the Aboriginal Dreamtime* (Rochester, VT: Inner Traditions, 1991). In the seventeenth century, the Japanese rejected guns and gunpowder for a period of two hundred years. http://infowarethics.org/gun.saga.in.1600s.japan.vs.germany.html.

brain to respond. The batter cannot think about a strategy to hit the ball. The ball comes faster than the reaction time of the brain.[6] Thus the batter has to commune—with his or her will—with the will of the pitcher. We watch baseball because we can participate in this communion in the will realm. Perhaps this applies to most sports and to their power to fascinate. We participate resonantly in more highly developed powers of will.

Technology, too, relates to deeds, to actions in the world, not to feelings or thoughts. Those come afterward, as reactions. Technology evidences the power of will. Thus our relationship to technology is as shrouded in ignorance as our relationship with our own will. When people say that technology is inevitable, they admit that they don't have control or much awareness of the realm of the will. However, the aborigines demonstrated that they knew their own will when they refused to take up complex technology.

Let's look at a fuller picture with each of the tools mentioned in this Image.

THE SHADOW OF THE DAGGER

The dagger misused can be destructive. Swords are turned into plowshares, and back again to swords—they are really the same.[7] The dagger becomes a tool that destroys soils, as witness the diminishment of topsoil in every country from overworking the land.[8]

In its role as device of measurement, it can give us the grid system of GPS, Google Earth, and EPIRB, so that one is never lost. One can always be defined as in a particular place determined by a number grid of rectilinear

---

6. A simple demonstration: http://www.exploratorium.edu/baseball/reactiontime .html.
7. Isaiah 2:4 speaks about beating swords into plowshares as a step toward peace. Adam made Cain into a plowman, who used the plow itself as a weapon to murder Abel. William Blake thinks so, as witness the shovel—a version of the plow—next to the dead body in Blake's print about Cain and Abel.
8. David Montgomery, *Dirt: The Erosion of Civilizations* (Berkeley: University of California Press, 2008).

references. Compare this to an indigenous view, where one relates to mountain and stream, marsh and sea, through personal knowledge of their curves. One is never lost because one relates to the living Earth.

Measuring all the land and dividing it up can divide a community, creating jealousy around possessions and distance between people. "Good fences make good neighbors" seems an admission of failure of a social network.[9]

THE SHADOW OF THE FIREBIRD

The firebird misused can become atom bombs…and arson, napalm, war in general with its emphasis on "firepower," and nuclear power plants with their "hot" water and occasional meltdowns.[10] The Fukushima meltdowns of 2011 illustrate that the firebird theme, stimulated by this Venus eclipse, is out of control.[11]

---

9. A medieval proverb used by Robert Frost in his poem "Mending Wall."
10. The pivotal experiment of the development of nuclear power, in which Leo Szilard and Enrico Fermi supervised the runaway chain reaction of radioactive materials, thus demonstrating the feasibility of nuclear bombs as well as nuclear power plants, occurred on December 2, 1942, in Chicago, with Venus at 20 degrees Scorpio 6 minutes, exactly opposite to the location of Venus at the eclipse of 2012. The pair of Images in Taurus/Scorpio at 20–21 degrees was active in that pivotal experiment. At the chain reaction experiment, the Sun lay at Antares, what Robert Schiappacasse and I have termed the Star of Death and Resurrection.
11. At the initial Fukushima explosion, at the time of the earthquake and tsunami, Venus lay at 17 Capricorn. Saturn lay at 20 degrees Virgo and 40 minutes, a trine of 120 degrees to 20–21 degrees of Taurus, showing a resonance with what lives in this Image. Fukushima now looks to outstrip the previous large dysfunctions of the firebird, at Three Mile Island (March 28, 1979) and Chernobyl (April 26, 1986). At Chernobyl, Saturn was at Antares (see previous endnote). At Three Mile Island, Saturn lay exactly square to the Aldebaran/Antares axis. The Royal Stars of Aldebaran (Eye of the Bull) and Antares (Heart of the Scorpion) are described elsewhere, especially in Tresemer, with Schiappacasse, *Star Wisdom & Rudolf Steiner*, chapter 1, and are an additional influence in this dance of the firebird with the cosmos. In an astrological sense, they are "close"—five degrees—to the 20–21 degree mark, close enough in the reading of a personal chart, but not close enough for the sharp indications that we seek in this research paper about the Venus eclipse.

The firebird is a favorite tool of the military. They have now created lasers of focused light so powerful that, when aimed up into clouds, it can trigger cascading energy effects that appear as intense storms that destroy the gentle earth of the enemy.

Earlier we mentioned that the firebird can be expressed as the electromagnetic powers of the microwave to cook food. However, when the firebird becomes rampant, it cooks things we don't want cooked. Those electromagnetic powers we use from the firebird to cook our food can also, when focused in a mobile phone held against the head, begin to cook the brain.

Recall the statement by Steiner, "The bird is the flying thought." When these thoughts are focused, they can accomplish deeds, first by warming the object of attention. When unfocused, they wreak havoc. Electromagnetic waves and electricity in all its forms can pollute a house or a whole city. Excess light can cut us off from the heavens by overpowering the twinkling stars.

The Shadow of the Cauldron

The cauldron misused can become enchantment with other places and times, and fascination with false realities—through television, computers, and all the other devices into which we stare and which take us away from our present surroundings.[12] Think of the story of Narcissus, pictured by the Greeks as a young man so infatuated with the reflection of his own image in a pool of water that he could not leave, and so died there, addicted to his own reflection. One might assume he was looking at the reflection of his own face,

---

12. Television was defined in the Oxford English Dictionary in 1907, when it was mostly imagined and not yet realized, as "the action of seeing by means of Hertzian waves or otherwise, what is existing or happening at a place concealed or distant from the observer's eyes," in other words, exactly as one would describe the magic cauldron. "Tele" means far off or distant.

but perhaps he was seeing all that the waters had held secret, and became enamored of other places and other times revealing themselves to him, so that his consciousness abandoned his physical body, that great vehicle which brings us repeatedly back to the present moment.

The cauldron can link us instantly to any place and any time we choose. "Consult Guru Google," we are told, or "Go to Wikipedia and you'll get your answer." People have come to expect that all can be known through these means, that there are no mysteries left beyond the magical surface of one's computer screen. The flood of images, facts, and opinions blocks one from self-initiated action, and one becomes atrophied in the will. The cauldron is always better at seeing; the individual feels small in comparison.

The modern cauldron—the television—illustrates an important aspect of technology: Not being human, technology does not sleep. It thus constitutes an attack on sleep. It is also an attack on nutrition because, like Narcissus, many watchers pine away unfed, enchanted by its images. Finally, it steals attention—one could even say sucks attention—leaving the person less full of life energy than before. Observe what happens when you view television (in its various forms, including movie theatres, and what you hold in your hand): Though the content seems to go into you, begin to notice what flows out from you to it.

We spoke about the electromagnetic fields around the cauldrons of water of old, seeing positive possibilities there. The electromagnetic fields around the cauldron's modern counterpart are so full of different kinds of information that they overwhelm anyone sensitive enough to hear them. Such people exist, who can hear all the stations of the radio, as well as police channels, and airline communications, and many of the other electromagnetic signals that are sent out every minute of every day. People at the edge of such sensitivity suppress that ability, as it overwhelms the sanctity of one's inner life. Thus the water's resonance and the firebird's cooking combine.

The rite of baptism, which we mentioned earlier, has a counterimage in waterboarding. In the earliest days of baptism, the body was held under water so that the person technically drowned, but just for a moment, as the spirit soared up into spirit realms. At just the right moment, the body was lifted up out of the water and revived. This was performed within a context of loving warmth, using that brief brush with death as a freeing of the spirit from identification with the necessities of the body, thereafter finding a love for the physical body as the vehicle in this world of the love that had been encountered in spiritual worlds. To use this same technique in a context of hatred, as in the waterboarding of criminals or of innocent people treated hatefully, pollutes the connection between spiritual worlds (what we knew before birth, and will again after death) and the mundane world.

The Shadow of the Ring of Power

When the livable surface of the Earth expands, it shifts. Either increasing the diameter of the Earth, or raising more earth up out of the sea means great rumbles and earthquakes. In the two years before and after the Venus eclipse of the Sun, will we observe this shadow effect?

The ring of power misused can become J. R. R. Tolkein's "ring," about which he wrote his famous trilogy, a ring indeed with too much power. A ring can become a symbol, as in the Pope's ring, which visitors are expected to kiss.[13] Think of all the emblems used by religions and nations to promote their ideologies, whose wars of words turn into wars with knives and fire(birds).

Consider the widespread use of emblems, logos, and pictures to promote oneself, as in Facebook, or one's personal website, or one's blog, or Twitter accounts. Magazines and their electronic counterparts now expect authors

---

13. The recent movie *Salt* pictures a man who trains children to do his bidding, including murder and mayhem, and to punctuate their deeds by kissing his ring. The ring seems to hold the power over their minds.

to write about their opinions and their lives for free in order to "build an audience based on your personal brand."[14] These promise relationship but are not relational. People talk to "walls," and not to each other. Shouting "Me, me, me!" does not bring on relationships.

Speaking so-called affirmations to the world—such as "Every day, and in every way, I am becoming better and better" or "The house at 700 14th Street will soon be mine"—is a kind of use of the ring of power, trying to impress your intentions into the world. The movie *The Secret* and its companion book became so successful because they promised success with this method. These techniques, which have been around for a hundred years, concentrate on material acquisition, a modern version of Jamshid's desire to expand the material of the Earth.[15]

All communication technology—including the old-fashioned telephone,[16] another version of the magic cauldron taking us across barriers of space and time—seems to take us away from the most important teaching for our soul's development, namely relationships. Thirty SMS texts a day does not assist relationship;[17] this activity redefines relationship as banal, impulsive, narcissistic, isolated one from another with a machine in between. Whenever we leave these messages on any of these media, we are a person speaking to a "wall"—or to a pool of water in a magic cauldron—a monologue masquerading as dialogue. We now seek to mimic the rampant impressions coming at us by trying to impress others.

---

14. David Carr at www.nytimes.com/2011/02/14/business/media/14carr.html.
15. Émile Coué introduced the frequent repetition "Every day ..." in 1920 in France. The naming of a house that you'd like is pictured in the film *The Secret*, based on the New Thought philosophy of Wallace Wattles's 1910 book, *The Science of Getting Rich* (Rochester, VT: Destiny Books, 2007). The Avatar system of training (www. avatarepc.com) offers a more sophisticated approach to formulating intention while dealing actively with obstacles, rather than sweeping those obstacles under the rug.
16. *Tele-*, distant; *-phone*, sound; thus sound from a distance.
17. The average now hovers around 50 per day for teenagers, and is rising quickly. See www.en.wikipedia.org/wiki/Text_messaging.

Genetically modified organisms (GMO) evidence a use of the ring of power. Scientists seek to control a particular pest by putting a pesticide into the genetic makeup of a plant. Many pests die, but not all, and resistant subspecies thrive. In the meantime, humans and animals consume the pesticide in the food, or in the honey made from the flowers of the engineered plant. And the genes of that plant spread to other plants not so engineered, a branding with the ring of power that expands outward, completely out of control.

Any mass thought-form, stimulated by the cleverest bearers of the ring, the advertising industry, can get out of control, and actually create the effects that it predicts. Thus prophecies can often fulfill themselves through mass attention, showing also the manipulation of the firebird. Perhaps December 21, 2012, will become an example of such a self-fulfilling prophecy by focusing mass attention on an event.

A final example is in our system of money. Exchange begins with barter of useful things or services. To make this easier, a currency was created, each piece of paper representing something useful, a good or a service—or, though not as good, yet something solid, a piece of gold. Fiat money, backed by nothing except reputation, shows the effects of the ring to grow and grow through rampant printing, disrupting the basic rite of interpersonal exchange of goods or services between human beings.

Technology and Semblances

The misuses of these tools and their descendents have served to disembody human beings, at a rate that is becoming epidemic. *World of Warcraft*, *Call of Duty: Black Ops*, and other electronic games engage the full attention of hundreds of thousands of people at every minute of the day.[18] The

---

18. Michele Foley, "How Many People Currently Play World of Warcraft?" *PC Magazine*, Dec. 27, 2010; www.geeksugar.com, October 8, 2010.

players' semblances, assumed personalities, or agents—arrogantly called "avatars"[19]—appear to be in relationship, yet they are not. Little thought goes to the consequences of killing in a bloody manner the agent of another person from some other place on the planet, who watches and hears himself or herself die by the hand of an unidentified but quite real other. Pause to consider this question once more: You enter into a cyberspace to control your semblance—generally a hunky, tattooed, weaponed warrior, regardless of your own physical manifestation—which encounters other semblances, also connected to living, breathing human beings somewhere else on the planet. You press your buttons and manipulate the machine that is you. You utter grunts, sneers, and interjections: "Hah! Got you!" You kill your opponent, or you are killed. The semblance dies in agony. You move on, killing others or being killed. Each one seems a chance encounter, and you may never meet that person again. You certainly don't know his or her name or hometown. But the energy body works differently. It knows that a connection has been made between two human beings, a nasty one. As you add points (number of kills) to your profile, what is happening at the level of the soul, which records all of these connections? Every day and night, thousands of people are murdered in cyber wars in this way. We have little awareness of the consequences to this training of our young to kill.[20]

---

19. Traditionally, "avatar" has meant the incarnation of a divinity from spirit into flesh. An incarnated divinity behaves divinely in the material realm.

20. See Dave Grossman, *On Killing: The Psychological Cost of Learning to Kill in War and Society* (New York: Little, Brown, 2009), and Dave Grossman and Gloria Degaetano, *Stop Teaching Our Kids to Kill : A Call to Action Against TV, Movie and Video Game Violence* (New York: Crown, 1999). The authors point out that throughout history soldiers have always had an aversion to shooting one another. Dave Grossman is a lieutenant colonel, so his assertion that the military has devised combat video games as a way to desensitize future soldiers has to be taken seriously. Brett Litz et al., "Moral injury and moral repair in war veterans: A preliminary model and intervention strategy," *Clinical Psychology Review* (29:8, December 2009, pp. 695–706) terms this a "moral injury" that has its worst consequences for soldiers returning from modern wars.

Confession in the Catholic Church can now be made by iPhone to an application (an "app") that responds as if it were a priest, but is not linked to a human being.

Cosmetics are a ubiquitous manipulation of semblance, including cosmetic surgery and injections of toxins. Young women evaluate themselves in relation to the models that are paraded before them by the image-makers. They disembody themselves by taking on a different appearance than that given by the wisdom of the body itself.[21] Botulinum toxin, or Botox, actually paralyzes the muscles so that wrinkles do not form. This in itself is a hint—living skin wrinkles over time. Wisdom comes from life, and it has wrinkles in it.

Separated from actual bodies, people expect to see instant effects from impulse, a push of a button that appears to operate a rifle, bazooka, magic lightning weapon, forgiving priest, or skin revitalizer. When the semblance of a deed is made more powerful, true will is diminished.

Technology is not, in fact, necessary to perfect one's soul nature. In the eyes of some, it may seem necessary to world evolution. In the eyes of others, it may indicate degeneration of the nobility of the human being. When we recognize that the greatest value is healthy development of individual and humanity, we can create a moral stance toward technology.

The need for this moral stance presses upon us now as some are predicting a moment when technology will become smarter and bigger than the human being and will gobble it up, making a kind of mixed creature, a cyborg that combines biology and technology.[22] We have to ask if we are ready to give away

---

21. The most potent demonstration of transformation of a normal person to an image of what people would like to see is given at http://www.youtube.com /watch?v=hibyAJOSW8U.

22. Raymond Kurzweil, *The Singularity Is Near: When Humans Transcend Biology* (New York: Penguin, 2006), enthusiastically predicts this event occurring in the near future. At Kurzweil's birth, Jupiter lay at 20 degrees and 9 minutes of Scorpio, thus exactly in the polarity energized by the Venus eclipse of 2012.

entirely the benefits of warm relationships in favor of becoming a mechanical tool. And who will control the mechanical part of us?

We have in this chapter looked at the shadow sides of the marvelous tools of Jamshid. We have seen that tools are amoral, and are not always used for the good of all. Though we may envy Jamshid's magical implements, we realize that modern human beings are actually surrounded by their counterparts today. And we are pressed to make serious choices about their use. Something new arises for this Image at 20–21 degrees of the Bull—its opposite. As the Sun and Venus connect in the Bull, directly opposite, through the Earth, another Image is unfolding in the Scorpion.

The polar opposite to the Image of Jamshid and his tools drives up the stakes in this matter, and also shows a way to come to terms with the shadow side of Jamshid's gifts.

HIGH STAKES FROM THE OPPOSITE

We have been looking at the shadow side of these implements, the opposites to the positive uses, and positive spin on the uses of these tools. We can also look at the opposite of the Image at 20–21 degrees of the Bull, namely the Image from 20–21 degrees of the Scorpion, because the full alignment at the Venus eclipse looks like this:

20–21 Taurus—Sun—Venus—Earth—20–21 Scorpio

This Image from 20–21 Scorpio will also be stimulated by the Venus eclipse of 2012 for some years to come:

However, this was from the point of view of the Sun (heliocentric), rather than from the point of view of the Earth (geocentric), the latter being our typical focus for this study. When something—an event or a birth—is related more to the Sun than to the Earth, it can indicate a larger framework of evolutionary change. Thus Kurzweil may be correct in his predictions, making the need for a moral stance to this change even more important.

> The dark streets are filled with the sick, the bleeding, the weak, the blind, the near-dead. Many were healed before and then relapsed. They wait in line, yearning for another chance. To the relapsed, the healer speaks: "The hands of your heart are withdrawn and locked up, for you are filled with darkness." The healer works in the night, bringing light into darkness, moving down the line along the street. The healer passes over some completely.

Picture how relapse works. One was healthy and then lost that health in a specific way—blindness as a failure of the eyes or bleeding as a failure of the skin—or in a general way—"sick," "weak," "near-dead." Already that challenges us with the enormity of suffering in the world. But "relapse" adds another part of the story. A person was healthy, then sick, sought help, and was healed. Imagine the feeling of being sick, then healthy again. Is there relief? Then something happened that the healer defines, a contraction of the hands of the heart, a withdrawal. The healed one holds on to the healing for himself or herself alone. Trying to lock up the light of health in one's own being, he or she creates only darkness. Illness returns, perhaps the same one, perhaps another form of illness. In a workshop on this pair of Images in Melbourne, we set up the scene whereby the ill were lifted up by a healer and approached the light, but then something happened—some turning point where the healed person's hands were withdrawn from the light of healing and contracted back into the self, disconnected from the world. The healed person turned his or her back to the light of healing and walked away, until the illness took over again.

We needn't consider this movement away from healing negatively. Indeed, we can all find examples in our lives where we've done just that. By concentrating on the self for a while, we integrate what we've learned, even if we cut ourselves off from the ministrations of others that sustain our lives. We are meant to come back, to rebalance, to return to health, wiser.

Everyone who has been ill knows that the body generously offers a second chance. In the Image, the once-sick and the twice-sick wait in line, yearning for a third chance. The healer generously gives another chance. Some are exceptions, those whom the healer passes over completely. Overwhelmed by the number of injured, the physician (in a process termed "triage" in the First World War) chooses who can be saved, and passes by those whom he judges cannot be revived. The sense of despair underlying this Image is tremendous.

We can think of the will as expressed through the hands and feet, arms and legs. The hint from this Image is that, for the greater king, deeds of the will must emanate from the heart, the home of feeling, and that this must be done freely with light, to light, and in light.

The darkness referred to—darkness in the heart and corresponding darkness in the streets—describes the world of the narcissist or the losing of oneself in exclusive self-reference. Even around others, these people are closed off, alone, despairing, not noticing that the world's ability to sustain life is suffering, as witness the dramatic rise in extinction of species.[23] The rise of social networks provides a measure not of social facility but of narcissism, making it achingly clear that true relationships are based on warmth and common activity rather than on image.

When inquiring about the use of the will, it helps to think through the use of hands. What do your hands do? Observe yourself objectively, not with judgment. What is your experience when you shake another's hand? How often does that happen? Do you feel the other person, look into his or her eyes? Or is the handshake perfunctory, or even something that you would like to get through quickly? What do your hands touch during your day? What do you imagine heart-originated hands would touch, or what

---

23. Species extinctions have gone exponential in the last few years, and this eclipse by Venus in this polarity (Taurus–Scorpio) may increase this effect or bring humanity's attention to it, or both.

would be the quality of that touch? Do you have someone who touches you, or whom you touch, with ease and familiarity, with warmth and light?

These two Images combined demand that you explore the quality of your relationships, and the quality of your hands touching and doing deeds in the world. Reacquaint yourself with your hands.[24]

A Moral Use of Techne

Techne is the goddess who rules technology, that is, the application of human cleverness and ingenuity to matter in order to create the artificial.[25] In our modern society we expend more and more time and energy making things that are useful and efficient, that promote the picture of kicking back with a drink in hand to relax in front of the television. We expend less and less time making something that is beautiful for its own sake, something that is noble, something that gives pure delight. When a craftsman fashions something from raw material, either from mirroring the world or from expressing inner experience, attending to the beauty of the creation rather than to its utility, a deep love is nurtured in the craftsman and in all who behold what has been made. As utility increases and art for the sake of beauty decreases, we may find that the technology of efficiency will grow in angry power and will confront our humanity with that anger. We may then discover that our sense of beauty and delight has disappeared. We may find ourselves within the picture of the Scorpio Image, *because of* decisions made in the context of the Taurus Image. Every day that we create something only utilitarian, or

---

24. Lila Sophia Tresemer and David Tresemer, *One Two ONE: A Guidebook to Conscious Partnerships, Weddings, and Rededication Ceremonies* (New York: Lantern, 2009), explore ways to empower relationship for personal development, including activity of the will, through many exercises to link hands, hearts, and concepts.

25. *Artificial* comes from *facere* (to make), and the word *art*, which means craftsman, and at its root means joining something together. Thus the foundation of art comes from imagination and execution of the joining of two pieces of wood that grew separately, held either by a dovetailed mortise and tenon or by pounded nails.

even use the technology of convenience without making an effort to balance it with beauty in our own lives, we press toward the Image at Scorpio in our own lives and in all human lives to follow.

How does Jamshid use the powers of technology? He prolongs life by hundreds of years for people and animals. He avoids the end of life by expanding the size of the Earth. He deepens Siddhartha's prison by dispelling old age and death. But how does death function in the world? It quickens; it stresses; it causes us to work the heart and to find the hands of the heart. A caring heart comes from having had cares. The heart becomes a purified heart. The hands can reach out. At death, the hands can become the wings of the heart.

Finding one's heart that can direct the will through the hands is no small matter. The words heart and soul occur frequently with little precision. Most people sense that something wishes to develop in the center of the chest, but aren't at all clear about what. Preparation for the heart as a strong and confident organ of love-in-action requires some attention. Hold your hands to your heart. Feel how it might grow in size, not physically, but energetically, to become a lighted mansion with many chambers—not a concept or a sweet idea, but something as strong as your leg muscles when running.

The Kings of Technology

Recently I spoke with someone who as a business consultant has worked closely with those responsible for the very rapid rise of computing technology—those who have taken computers from little helpers in the accounting department (faster than abacuses) or an evening's entertainment once a week (the ten-inch black-and-white television sets of the 1950s) to an activity that takes up much of most people's time. My client described those in charge of the development of this new technology as having very undeveloped social skills. They had little personal warmth or empathy;

they shrank from conversation, social encounters, and even eye contact; they excelled at focusing on dazzling lights that blinked and twittered. She summarized her experience of these people by diagnosing them with Asperger's Syndrome, whose symptoms she listed. Asperger's is thought perhaps to be its own syndrome, or perhaps a form of autism, the root *aut-*, meaning to oneself, that is, narcissistic.[26] She concluded that "they were unable to play like other children, so wanted to make other children play like them"—in other words, alone with the flickering firebird playing across the face of the image-creating cauldron, imprinted with the ring of branding, as a dagger cutting them off from relations with the Earth and with other warm human beings.

This caused me for the first time to inquire, "What kind of person was King Jamshid?" I had vaguely assumed that, as king, he had the best in mind for his people. But perhaps he was self-possessed, interested mostly in playing with his magical toys, to the extent that he didn't notice that the world was suffering, and had to be reminded by a divinity that things were not going well. One often assumes that philanthropists are what the term implies: lovers, *philo-*, of humanity, *anthropos*. But the world abounds with examples of sociopaths and egoists who give money to charities. Perhaps Jamshid was such a character, and that answers the question of why the Image goes to the next step, a greater king.

---

26. I am not referring to present discussions on the Internet about whether or not these leaders of computing technology ought to be diagnosed as having Asperger's Syndrome. I am relying on the firsthand experience of one who worked with them daily, an intelligent person, though not a psychologist, trying to make sense of the leaders of the computer revolution.

4

GREATER KINGS AND QUEENS

In the Image for 20–21 Taurus, a wise visitor reminds the seven philosophers that there is a greater king. Beyond Jamshid, the king of great powers, a greater ruler exists. The wise visitor doesn't announce this new king. He reminds the seven of something they already know.

The greater king does not rely on external tools. Why not? People collect rings, necklaces, bracelets, amulets, talismans, crystals, chalices, swords, lucky coins, magic wands, all versions of what Jamshid had. At a certain point, it all seems so much to care for, to ensure the appropriate use of each, to clean and tend the myriad pieces. Fewer tools means that one relies on one's own body and being as the tool for one's divine intention. Having one's tools or capacities held in the heart means that they are always ready and always used appropriately.

GREATER KINGS—THE BODHISATTVAS

The seven philosophers wonder about the greater kings. Could it be Zarathustra? Could it be Melchizedek? Could it be Inanna? Could it be Solomon?

This sounds like a list of all the greats. We could include the names of powerful earthly kings, such as Nebuchadnezzar. We could include the names of powerful kings of the spirit, such as Jeshua ben Pandira or Pythagoras or Hermes or Jesus Christ. We could include names of kings both spiritual and societal, Solomon or Abraham. We could include queens of societies and of the spirit, Inanna or Deborah or Sheba. We could include the names of Ascended Masters, such as Koot Hoomi or El Morya or the seven Rishis. We could add to Zarathustra the names of other Bodhisattvas, such as Hermes or Buddha or Scythianos.[1] We seek to find among the leaders of humanity those who exert a power of great vision, and who rely less on exterior tools than on their own capacities, held in the heart.

The seminal study of star wisdom and culture, *Hamlet's Mill*, identifies Jamshid with Saturn, both as planet and as Kronos, Lord of Time, Lord of the Golden Age, and, as we might learn from Rudolf Steiner, Lord of Cosmic Memory.[2] All leaders have a streak of Saturn in them. Connected with the Sun, and thus greater powers of heart, as strongly urged in the Image opposite at Scorpio, this Saturn-infused leader becomes great, a great king.

In reading the history of these characters, we realize that we actually know very little about them. Groups have coopted them for their particular propaganda purposes. Lacking even their birthdates, we can only truly know them by inner reflection. Let's examine a few of these figures now.

Zarathustra lived either in the sixth century BCE in Babylon, or in the sixth millennium BCE in the Ancient Persian age, or both, with other incarnations in between.[3] We know of no implements that he used, except perhaps, as the Zoroastrians claim, the baresman or barson, a collection of thirty-five twigs of tamarisk, pomegranate, and other trees, to affirm his gratitude for

---

1. The website www.cosmochristosophy.org takes up the question of masters in contrast to bodhisattvas, examining many individuals critically. T. H. Meyer and Elisabeth Vreede delve deeply into this in *The Bodhisattva Question: Krishnamurti, Rudolf Steiner, Valentin Tomberg, and the Mystery of the Twentieth-Century Master* (London: Temple Lodge, 2010).
2. Re Jamshid and Saturn, *Hamlet's Mill*, op. cit., 146–147.
3. The names *Zarathustra* or *Zoroaster* or *Zaratos* can be seen as successors in this line, or reincarnations of the same individuality. Andrew Wellburn, *The Book with Fourteen Seals* (London: Rudolf Steiner Press, 1991). "BCE" = Before the Common Era, the modern version of "BC," Before Christ. Instead of AD (*Anno domini*, "the year of our Lord"), we use here "CE" (Common Era).

the natural world.[4] He was the wisest human being alive, and had a personal experience of the beings of the heavens—in the so-called first hierarchy of Thrones, Cherubim, and Seraphim—from which he suggested a map of the heavens, the zodiac as we know it today. He experienced these as living beings, not as clumps of stars, or connected dots, or districts of the astronomers' zig-zag map shown at the end of Appendix C. He remains the main inspiration for the discipline of astrosophy, star wisdom. The map according to the star-based tradition of Zarathustra of the Venus eclipse is given in Appendix D.[5]

Melchizedek—king or great (*melchi-*) priest or righteous one (*-zedek*)— gave the Grail cup to Abraham, who passed it on to Yeshua. Melchizedek was the mysterious desert king and priest who brought the ceremony of bread and wine to Abraham.[6] The Holy Grail can be seen as a kind of magic cauldron, its tradition passed down from Jamshid's time.

Abraham took the ceremonial instruction and the implements and spread them to the world. He was also known for his use of the ritual knife in nearly slitting the throat of his son Isaac on the pyre of sacrifice, before being dissuaded by an angel.[7]

---

4. Zoroastrians claim that Zoroaster, or Zarathustra (perhaps also Zostrianos, as in Marvin Meyer, ed., *The Nag Hammadi Scriptures* [San Francisco: Harper, 2007]), originated the use of the *kushti*, a gathering of 72 fine white wool threads that they wrap around the body three times, performing this ceremony several times a day. A five-pointed pentagram has many internal angles of 72 degrees, linking 72 with 5. See Appendix A on Venus and the number 5.

5. Zarathustra gave the equal-length zodiac of twelve beings, each with thirty qualities, making 360 different degrees in the heavens. From this we get all of our geometry (based on 360 degrees in a circle) and time (based on division into twelve zones, each with sixty minutes, which is twice thirty). The names of the first hierarchy of divine beings—Thrones, Cherubim, and Seraphim—denote the most developed of the nine ranks of angelic presences, the lower being archangels, and finally angels.

6. Genesis 14:18–20, also Psalm 110:4. Paul's *Letter to the Hebrews* stated that Jesus was a priest in this line of Melchizedek, 5:6, 10, 6:20, and all of chapter 7.

7. Abraham lived approximately 175 years, some say from 1812 BCE to 1637 BCE, some claiming his birth was 1946 BCE; in either case four millennia before the present, and at the beginning of the Age of Aries (vernal point entering Aries).

Inanna, Queen of Heaven and Earth, was celebrated by the Sumerian culture from thousands of years ago. Her story has recently been translated from cuneiform tablets found in Iraq.[8] I have coauthored a drama called *The Great Below* that presents her story in relation to modern times.[9] Like Jamshid, Inanna received many gifts from the gifting God Enki, skills and qualities of personality that made her the apotheosis of all humanity. In the land of the dead, she was revived by bread and water, the resurrecting water presumably being carried in a Grail cup of some kind.

The Grail cup is common to many of those named. Even Solomon had an immense container for sacred water, called the Brazen Sea, that measured ten cubits across and five cubits deep (a cubit being the measure from elbow to outstretched fingers, approximately 20 inches). Its capacity was 24,000 gallons. He also commissioned many sacred vessels for his temple made of orichalcum, an unknown alloy of high value, perhaps a mixture of gold, copper, and silver.

THE TOOLS OF THE BODHISATTVAS

The Grail cup of Christian legend has many similarities to Jamshid's cauldron. Wagner's opera *Parsifal* pictures the Grail cup, brought from the crucifixion where it collected the blood falling from the body of Christ Jesus to the remote forest where it is protected by the Grail Knights. The Grail cup is the pivot around which the entire drama takes place, the tool that keeps the band of Grail Knights alive, though barely.[10] *Parsifal* was first performed on July 26, 1882, with Venus at 19 degrees and 2 minutes of Leo, thus square to the 20-degree mark of the Venus eclipse, in the year of the previous Venus

---

8.   Diane Wolkstein and Samuel Kramer, *Inanna, Queen of Heaven and Earth: Her Stories and Hymns from Sumer* (New York: Harper, 1983).

9.   *The Great Below*, from DavidAndLilaTresemer.com.

10. For more on this topic, please see chapter 8, which tells the story of Parzival in greater detail.

eclipse, namely 1882. In the opera *Parsifal*, the lance that wounded the Fisher King—paralleling the golden dagger of Jamshid, a tool that can be used for good or ill—is stolen. By extension, one wonders if Jamshid's dagger also has the blood of spirit on it. In Wagner's opera, a firebird as dove of light appears in the opening scene, when the Grail is uncovered.[11]

Trevor Ravenscroft popularized the Grail and lance in his books, *The Spear of Destiny* and its sequel, finding that the power of these divine tools—parallels to Jamshid's—was still operative after many centuries.[12] We shall return to Parzival in chapter 8 because of the many ways in which the characters associated with this tale relate to the Venus eclipse event.

The little bits known about these leaders don't give an entire picture, yet they can be used as gateways to come to know them better through inner contacts.

THE INSANITY OF KINGS AND QUEENS

How can one come to know such a leader? I would like to investigate Melchizedek in greater detail. Experts say he was so ancient that we can't know much about him. But Paul identified Jesus Christ as a member of the Order of Melchizedek, something that has completely mystified commentators.[13] Was Melchizedek alive at the time of Abraham and at the time of Jesus also? And there are tales of Melchizedek being seen in more modern times. This alerts us to someone who, "without father or mother, without genealogy, without beginning of days or end of life, like the Son of God

---

11.   Parsifal is Wagner's spelling of Parzival. As detailed in chapters 7 and 8, Wagner's birth shows a prominent connection between Venus and the Sun.

12.   Trevor Ravenscroft, born on April 23, 1921, with Sun (9 Aries) conjunct Venus (8 Aries), is thus related to the Sun–Venus connection that we've been following. *The Spear of Destiny: The Occult Power Behind the Spear Which Pierced the Side of Christ* (York Beach ME: Weiser, 1982) and *The Mark of the Beast: The Continuing Story of the Spear of Destiny* (York Beach ME: Weiser, 1997).

13.   See footnote on page 41.

he remains a priest forever."[14] John Lash has daringly pointed to this and other references to suggest that this great king was not human, that he was, indeed, one of the deathless spirits—the Archons named in the ancient scriptures of the second-century manuscripts found secreted in a jar at Nag Hammadi, Egypt. The Archons were created as a side-effect when the divine goddess Sophia became the body of this Earth. Rudolf Steiner would call them Ahrimanic demons or, in the case of Melchizedek, an arch-Ahrimanic demon.[15] The depth of cosmological understanding necessary to compre-hend Lash's presentation is beyond what we can go into here. It need only be said that he recommends that we question the motivations and actions of leaders. We look at the deeds of kings and queens (in their guises as presi-dents, religious leaders, business tycoons, and military brass) and wonder, "Have they gone insane? Don't they see that their actions are making things much, much worse?" And we can ask to what extent they may be influenced by retarding or regressive influences—a nicer term than the one that scares most people away; that is, demons.

Lash might question whether Jamshid was an Archon, or was influenced by one, in his insane desire to keep every human and every animal from dying. He challenges us to see that most technology is the attempt to control, squelch, and possess the human soul and human heart. Even the technol-ogy that looks promising—nanotechnology to cure disease, or thousands of video cameras in cities to ensure urban safety, or any invention or "break-through" that you can think of—has the consequence of reducing human

---

14. Hebrews 7:3.

15. Steiner adopted the name of the Persian lord of the underworld as representative of the dark and dense realms of hell. See John Lash, *Not in His Image: Gnostic Vision, Sacred Ecology, and the Future of Belief* (White River Junction, VT: Chelsea Green, 2006), his reading of the full extent of the Nag Hammadi texts (Marvin Meyer, op. cit.), as well as in his several YouTube posts, his interview with Jay Weidner, *Sophia Rising: Return to Planetary Tantra* (SacredMysteries.com, 2009), and his newer interviews at www.MarionInstitute.org.

freedom to develop the heart, which is humanity's true purpose in relation to the divine feminine origin of Earth. Neither longevity nor riches is the point of life, nor are these the proofs of goodness. Relation to the heart of the world, to Sophia, our mother, our genetrix, the one who is dreaming us, with all of our heart, our soul, and our mind—that's a vastly more worthy goal.[16]

As for the mystical mighty Melchizedek, we cannot confirm his purpose or his lineage. We simply have to ask the questions and observe, then from a distance of space and time feeling what's true about him. It's safer to differentiate between a public leader, one who wields power over people or soldiers or money, and a private leader, one who has devoted himself or herself to the service of life, what we can call a bodhisattva.

THE MEANING OF A BODHISATTVA

What do bodhisattvas look like? They have an interest and capacity for care for others. They have no tradition. Typically they are empowered in their work past thirty years of age, and are not recognized before that time. They are completely devoted to the spiritual maturation of human beings, and indeed of consciousness at all levels, down even to the mineral. Rudolf Steiner's book *Turning Points in Spiritual History* is about such people and their impact on human society. The lives of these human beings are all turning points for human development. Steiner gives a picture of twelve such bodhisattvas, some incarnate and others not. We can learn from them, even at a distance of space and time. The Venus eclipse Image suggests several, and does not give us certainty about who, in fact, is the greater king. That confusion does not cripple us. Each of the ones named can become the mentor that works with an individual to guide through further development.

---

16. Luke 10:27: "Love the divinity to whom you are related with all your heart, all your soul, all your strength, and all your mind."

These individualities, living even though not in body, can guide one through challenging times, such as we now experience in the world.

The beginning of the Image empowered by Venus crossing the face of the Sun lists magical tools. The latter part of the Image lists powerful guides of humanity. Going from the first to the second part marks a step in evolution. And there is a further step.

The Greater King

What rules over the three centers of thinking, feeling, and willing? What rules over the seven energy centers of the human body? The first body or entity to rule over them is the personality. But there is a greater King, the "I AM," the spark of the divine that is not a gift of heaven to humanity, but rather an essence of heaven that *is* humanity. The "I AM" is aware of itself: "I exist!" "I perceive!" "I am aware of the world, and I am aware that I am aware!" When you speak those words, the one speaking in the most pure and comprehensive way is a truer you. No matter what skills you have in any of the above three or seven centers of power, a greater vision spans them all and then some. Thus this Image has in it a grand drama of hierarchy, finding above the more obvious *things*, even if powerful things or spiritual things, a principle or being that rules them all. The personality (the small "ego," the grasping "me-me-me!") seeks physical immortality; the larger "I AM" gazes over lifetimes, and does not avoid death.

While one must rely on the mentoring from elders, including from the great kings mentioned in the Image from Taurus, in the process of personal development, this leads to a sense of one's own sovereignty in close relationship with all of living creation. Sovereignty means realization of the royal self. It does not mean isolation and aloneness, but rather the ultimate insertion of oneself—one's Self—into the divinity that exists at all levels of being right here and right now.

Thus this Image relates a process of growth, from the egocentricity that comes from owning tools of great power, to devoted discipleship to powerful kings and queens, to devotion to wise kings and queens, to devotion to highest divinities of love and light, and finally to a sense of the regality of one's own nature which, when fully developed, is a divine principle. You know you've accessed that level of kingship or queenship when time and space melt before your vision, and when your only desire is to serve the good of the whole, the evolution into love of all of the world—when your hands flow spontaneously from your heart in many deeds great and small.

5

STAR SISTERS, STAR BROTHERS, AND EVENTS

The first breath of each person born through a given stargate—that is, with the Sun at a particular degree—pulses with the themes of the Image related to that degree in the Oracle of the Solar Cross. We call those who enter the world through the same stargate star brothers or star sisters. Looking at a few of the star brothers and sisters for 20–21 degrees of the Bull can instruct us as to the main themes at work during the Venus eclipse of 2012.

CAGLIOSTRO

The eighteenth-century magical genius Alessandro Cagliostro was born with the Sun in this degree. He was called Grand Master by his admirers and numbered kings and queens among his clients. He was also hated by powerful people, including Goethe and the Pope. He healed thousands with magical powers and magical implements, and had an unending source of wealth that people thought he created with divine help through the magic art of alchemy. Those visiting his "Temple of Egypt" in Paris were first met by two servants clothed as Egyptian slaves, then by Cagliostro dressed in a black silk robe with an Arabian turban made of gold cloth and sparkling with jewels. On a black marble slab was the "Universal Prayer" by Alexander Pope, a man who also drew his first breath on this Solar Day.[1] Cagliostro

---

1. Alexander Pope, born May 21, 1688 (Julian calendar).

attempted to revive the ancient arts of Masonry, which go back at least to Egypt and perhaps earlier. Here was a king like Jamshid, with powers and magical implements.[2]

Was his magic derived from heavenly sources or from old lineages of black sorcery? We could say that Cagliostro enjoyed the earlier part of his birth Image but did not mature to the latter part of it.

The Inquisition didn't burn him, as burning ended in 1600. But they were happy to imprison people for long periods of time in dank and dark places, creating the kinds of illness that we meet in the Scorpio Image. Cagliostro died in this miserable manner.[3]

PHILIP II OF SPAIN

Philip II of Spain ruled the largest empire ever assembled in this world, owning large pieces of several continents.[4] The accounts of his life range from the Black Legend to the White Legend. The Black Legend depicts Philip as a terrible monster, destroying all that he met in Europe and the New World. This King Philip pressed for more and yet more gold from the enslaved miners in the New World, expanded the Inquisition to become a secret police force that imprisoned and killed political dissidents, and sent the sailors of his Armada to death at the hands of the English. The White Legend depicts him as a prudent and pious statesman, champion of the arts, leader of Spain at the height of its cultural development. In Philip II we can find an increase

---

2.   Iain McCalman, *The Last Alchemist: Count Cagliostro, Master of Magic in the Age of Reason* (New York: HarperCollins, 2003); H. P. Blavatsky affirmed that his powers were real. Cagliostro (b. June 2, 1743) was consider by many the author of *La très sainte Trinosophie* (*The Most Holy Trinosophia*), a lavishly illustrated book that promises magical access to the threefold nature of the divine feminine, Sophia.

3.   Cagliostro's Sun Image was in Taurus. His Earth Image (as explained in Tresemer and Schiappacasse, *Star Wisdom & Rudolf Steiner*, op. cit.) was directly opposite, in Scorpio. Thus he was pulled by his Earth Image at the end of his life.

4.   Philip II was born on May 21, 1527 (J).

in reliance on techne, and the improvement of convenience for all his people, especially the wealthy. We see also the negative, indeed dark, side for others.

LORCA

Federico Garcia Lorca responded to the entire Image of 20–21 degrees of the Bull.[5] He relied on fewer and fewer props over the course of his career, eventually summarizing the greatest gift that one could have and use as *duende*. Duende cannot be defined, only felt. It includes capacities of spontaneity, great power that rises up from the feet, engulfing the artist and then the entire audience. One could call it charisma, yet it is far more Dionysian, dangerous, and consuming. Lorca learned from his birth Image to eschew tools and to hold in his heart something great and wild. To Lorca, Jamshid's dagger spelled danger. Here is a piece from his play *Blood Wedding*, in which he deals with the tradition of the dagger in the Image into which he was born:

> BRIDEGROOM: I'll eat grapes. Give me a knife.
> MOTHER: And why?
> BRIDEGROOM: To cut them...
> MOTHER: (*muttering*) Knives, knives.... Curse them all, and the wretch who invented them....
> BRIDEGROOM: Let's change the subject.
> MOTHER: And shotguns, and pistols, and little razors, and even hoes and winnowing hooks.
> BRIDEGROOM: Fine.
> MOTHER: Whatever can cut through a man's body, a lovely man, in the flower of his life, who is off to the vines or the olives, because they are his, his family's....
> BRIDEGROOM: (*Lowering his head*) You've missed the point.
> MOTHER: ...and he doesn't return. Or if he does return it's so we can lay a palm leaf or a big plate of salt on him so the body won't swell. I don't know how you can carry a knife about you, or why I have these serpent's teeth in my kitchen.

---

5. Born June 5, 1898.

The mother's premonitions prove accurate. At the end of the play, the bridegroom stabs and is stabbed by his new wife's old lover. They both die by the knife. Note that the mother extends the knife to shotguns, pistols, little razors, hoes, and winnowing hooks—all tools that come from Jamshid's golden dagger.

Lorca fired the imagination in a way that needed no tools. Here is a passage from his "Ode to Whitman":

> Sleep on, nothing remains.
> Dancing walls stir the prairies
> and America drowns itself in machinery and lament.
> I want the powerful air from the deepest night
> to blow away flowers and inscriptions from the arch where you sleep,
> and a black child to inform the gold-craving whites
> that the kingdom of grain has arrived.

So few words, yet so rich! "America drowns itself in machinery" shows his fundamental distrust of Jamshid's magic implements, so many that the air is lost. One cannot breathe and drowns; a whole country drowns. Lorca is a positive example of the evolution suggested in the Image, grappling with the tools, then giving them up in preference to radical presence, to *duende*.

EVENTS—WARS

World events can indicate how an Image from the cosmos works into culture. Two such events have occurred in the degrees of this Taurus–Scorpio polarity, exactly opposite bookends for an immense conflict: Pearl Harbor and D-Day. The destruction of the U.S. naval fleet at Pearl Harbor took place on December 7, 1941, Sun at 21 degrees of Scorpio. The Japanese called it Operation Tora, meaning tiger.[6] D-Day, called Operation Overlord (con-

---

6. The planes were launched at 6:05 a.m., when the Sun was at 21 degrees, 16 angular minutes of Scorpio, thus into the next degree, though less than 30 angular minutes from the Venus eclipse. The first strike was at 7:51 a.m. Angular minutes refers to

necting to the notion of a powerful king) took place on June 6, 1944, Sun at 21 degrees of Taurus. It was the largest sealift in human history.[7] Both of these events are marvels of technology.[8]

This pair of events attunes us to power, the kinds of power that exist and how they may be used. We have spoken of turning points in people. These are turning points in history. These events hit home at a societal and world level. Exact opposites are often related and useful in the Oracle of the Solar Cross.

When people are lifted from the Earth prematurely because of an earthquake or volcanic eruption, Rudolf Steiner pictured them as working through what would have been the normal course of their lives in service of angels who are assisting humanity to evolve. This is a powerful picture that defies the term "accident." You can verify this assertion for yourself by feeling into it, by testing it with your intuition, perhaps in contact with someone you know who went through an untimely death.[9] Even people killed unexpectedly, as in Pearl Harbor, can be seen as completing the term of their lives in service of angels.

Wars are somewhat different. In war, what kills people is not an "act of God" or an accident. What kills people are other people desperately seeking not to be killed themselves. Aggressors running up a beach meeting other

a portion of a celestial degree, as opposed to clock minutes. The Earth rotates by fifteen angular minutes for every clock minute.

7. When the preliminary air strikes began, it was just after midnight, with the Sun at 21 degrees and 7 angular minutes of Scorpio. The landing on the beach began at 5:30 a.m. Though this is just into the next degree, it is 17 angular minutes from the Venus eclipse position.

8. One more event from World War II, related not to these degrees but to a Sun–Venus conjunction: Germany invaded Poland because it was claimed to belong to the Fatherland, September 1, 1939, Venus 12 Leo 38, Sun 13 Leo 55. This also was a massive display of superior military technology as the tanks rolled into Poland.

9. While Rudolf Steiner suggested that suicide was not in service of the angels, a person who has committed suicide is nevertheless an individuality with whom you can commune, and who may be living out the normal course of his or her life.

aggressors running down a beach, as in D-Day, are not unprepared, and have honed their hatred to a high pitch. The conflict can be seen as stimulated by the theme of the Taurus–Scorpio polarity—a clash of technologies, and observation of the dearth of hands of the heart. The deaths of Pearl Harbor and of D-Day, though different in tone, would weigh heavily on anyone who is born into these degrees—or anyone who chooses to experience the Venus eclipse in that degree in 2012. One can deal with this head-on, acknowledging the events and the suffering in the events as two groups exercise their technologies on each other, then offering a blessing from one's heart and an act of kindness through one's hands, and finally going past, to the deeper gifts of this polarity.

A special case arises from cyber warfare, where many people—mostly young people—staring into Jamshid's cauldron of the computer screen, actively engineer the murder of many people every day. The murdered and the murderers, though apparently unhurt in the physical, nonetheless accumulate a kind of baggage of hurt, which has to be dealt with at some point, either in this world or in the next. Though this happens every day, and is not related as an event to the eclipse of Venus, we can expect that the issues of technology stimulated by the Venus eclipse will see this kind of activity increase—or experience a reflective evaluation of its insidiously destructive nature.

EVENTS—THE NEUTRON CHAIN REACTION AS PRELUDE TO THE BIGGEST TOOL OF ALL

With Venus at 20 degrees of Scorpio, Leo Szilard and Enrico Fermi undertook the great experiment in 1942 that proved an atom bomb possible. Neutrons were bombarded until a runaway chain reaction began, which the scientists were then luckily able to bring to a halt.[10] This illustrates our

---

10. The Neutron Chain-Reaction experiment was undertaken by Leo Szilard and

theme about runaway technology and its uses for good or ill. Characteristic of the human penchant to experiment beyond the limits of control, the scientists attending this experiment did not know whether they would in fact be able to stop the reaction, and thus the whole world might have exploded, Jamshid's firebird run amok. That meeting with the firebird gave us the atomic age, from which we are suffering now, most recently as the threat of suitcase bombs deployed by shadowy "terrorists" against innocent people.

EVENTS—TECHNOLOGY TO THE MOON

With Venus at 21 degrees of Taurus, Neil Armstrong was the first human being to set foot on the Moon, also an astonishing feat of technology.[11] Did it really happen? Or was it the magic of the cauldron, giving the world false pictures for the purpose of political gain, as suggested by some?[12] Judging from the interviews with the astronauts, it was a life-transforming experience that actually occurred.[13] Edgar Mitchell was so impressed that he cofounded the Noetic Institute, which brings scientists and spiritualists together. No matter what actually happened, this event and the imagination

---

Enrico Fermi at the University of Chicago, on December 2, 1942, with Venus at 20 degrees and 10 minutes of Scorpio. We come back to Szilard's birth in relation to Venus in chapter 7.

11. July 20, 1969, Venus at 21 degrees and 2 minutes of Taurus.

12. There are many books and DVDs on the Moon landing as "hoax," the most interesting being Jay Weidner's *Kubrick's Odyssey* (www.sacredmysteries.com), wherein he demonstrates how Stanley Kubrick might have filmed the Moon landing with a silver screen backdrop (similar to a "green screen" system used today)—with Jay Weidner's important proviso that, though what we saw on television was faked, it was to divert the attention of competing countries away from the real technology that was used to visit the Moon and bring back Moon rocks and wide-eyed astronauts.

13. The interviews with Edgar Mitchell and others are completely convincing. Based on his experiences in the Moon landing, Edgar Mitchell went on to found the Institute for Noetic Sciences.

of this event put us squarely into the demonstration of technology and its uses—Jamshid's dagger became transport to the Moon.

The men, women, and events connected to the 21st degree of Taurus–Scorpio give us a sense of the fruits of these encounters with that degree. Through this study, we create a matrix of meaning which we can expect to be stimulated by the Venus eclipse of the Sun.

6

MENTORS AND SPOILERS

When you die, you lift up the harvest of your life to the heavens. These patterns that you've created in your life become amplified by the Sun and impress them upon the cosmos. The death occurs through a stargate, in a particular degree, where the Sun lies in relation to the background of the zodiac. Someone taking his or her first breath through that same gate can be affected by the patterns left there by those who took their last breath in that gate. In *Star Wisdom & Rudolf Steiner*, I called them bodes, as they either bode well and become mentors, or they bode ill and become spoilers.

Usually we're interested in the deaths of powerful individuals, whom we would expect to leave a strong impression in the heavens. We're also interested in mass deaths (such as in the two main world events occurring in this degree, Pearl Harbor and D-Day), whose many fatalities in the same degree can create a kind of haze obscuring the import of an Image coming from the stars and amplified by the Sun.

Some powerful person died through the gate through which you were later born. You passed right by him or her when you were born. You had your awareness on other matters, getting through the greatest trauma of your life. However, you did pass by these influences, and we have found that they have had an impact—and could have a greater impact if recognized and worked with actively. StarWisdom.org sends out mentor connections to people who inquire. To get lined up with the right mentors requires some calculation, as the calendar dates are unreliable after fifty years or so (owing to the "precession of the equinoxes," other calendar systems, etc.).

Once hooked up with a mentor, we suggest that you arrange an imagined tea with that person. We go through this process in some detail in the Tea with Your Mentor process at StarWisdom.org.

Usually a mentor–client relationship is set up based on the client's birth details. However, a rare phenomenon such as the Venus eclipse of the Sun becomes important for everyone. Here are the mentors for the event of the middle of 2012.

Mozart

Wolfgang Amadeus Mozart died in this degree. In *The Magic Flute*, Mozart gave us the character Sarastro—a representative in name and in power of Zarathustra, the leader of the school of astrosophy. In Ingmar Bergman's film of *The Magic Flute*, we see Sarastro in his spare moments reading an ancient copy of Eschenbach's *Parzival*, a key connection introduced in chapter 7 and developed in chapter 8.

I am very fond of Mozart, *The Magic Flute*, and the depiction of Sarastro. We can gain strength from Mozart's later music, that which was fresh in his mind and heart when he offered his life's deeds as harvest at his death. He can thus become a mentor for working with the themes of this Image of the Solar Cross.[1] Cultivate a relationship with Mozart's music, enjoyed with live musicians when you can, both before and after the Venus eclipse, to gain strength and insight into our times. Music becomes the magic formed from technology—the flute—that, when played with one's active breath while walking forward with one's beloved, leads the player and his partner through harsh initiations.[2] To make the flute work, one has to possess the

---

1. Died December 5, 1791, Sun at the edge of 21 Scorpio, the exact time of death not known (some reports place it at midnight, while others say some minutes later), in any case straddling the 21st degree of Scorpio.
2. The initiation themes of *The Magic Flute* are well researched in *The Magic Flute Unveiled: Esoteric Symbolism in Mozart's Masonic Opera*, by Jacques Chailley

qualifications noticed by the priests: "His heart is bold, and pure his mind." We encounter Mozart more intimately at the end of this chapter.

DEATH OF KINGS

Two kings died into this degree, Ronald Reagan and King Henry VI of England.[3] Both were mad toward the end of their lives—Reagan with Alzheimer's, whose traces could be seen even while he was still in office, and King Henry VI sometimes catatonic and sometimes hysterically singing while battles raged close by. The historians tell us that Henry was murdered while at prayer, though Shakespeare had him stabbed to death in an argument.[4] Either way, when men and women of power die by murder, it adds disturbances to our ability to feel the pure vibration of the degree.

We can perhaps feel the legacy of Ronald Reagan as a conditioning of our experience of this Image. Recently more facts about his presidency have come to light. Even though he had credible evidence from his own staff that the Soviet military was in decline and had few serious weapons, he revived the Cold War, labeling the Soviet Union the evil empire and increasing fear in the American public to the extent that his weaponization of space ("Star Wars," or Strategic Defense Initiative) almost came to pass. He created evidence for his claims and pressured Congress to increase mightily the expenditures on military weapons, as well as

---

(Rochester, VT: Inner Traditions, 1992). A fine guide to using Mozart's music to prepare for difficult times is Don Campbell's *The Mozart Effect: Tapping the Power of Music to Heal the Body, Strengthen the Mind, and Unlock the Creative Spirit* (New York: HarperCollins, 2001), and his new book, *Healing at the Speed of Sound: How What We Hear Transforms Our Brains and Our Lives* (New York: Hudson Street, 2011).
3. Ronald Reagan died June 5, 2004, and King Henry VI of England died May 21, 1471 (J), both with the Sun at this degree of Taurus.
4. Shakespeare wrote three plays about Henry VI and was sympathetic to the challenges that Henry faced.

convincing many other countries to weaponize. This is a president who was enamored of Jamshid's technologies and their potential uses for war.

ROBERT KENNEDY

Robert Kennedy was shot and descended into a coma in this degree.[5] Here is a man who championed the voice of the people and their needs, a man who was able to reach out from his heart, murdered in this degree. This disturbance must be acknowledged and then intentionally set aside in order to establish a pure, strong connection to the Image.

HARTMANN

Eduard von Hartmann, who earned the label "the philosopher of pessimism," died with the Sun in this degree.[6] Rudolf Steiner met Hartmann before publishing his own book, *The Philosophy of Freedom*, and was disappointed at his insistence that thinking could only wander in illusion, and that freedom was merely one of those illusions. One has to encounter and pass through the coloration of Hartmann's pessimism in order to find the true power of this Image.

TINTORETTO

Tintoretto died here. One could view his last great painting, his *Paradise*, the largest classical painting ever painted, at 74 feet by 30 feet, for inspiration to approach the Image. Or seek out any of his other paintings. Tintoretto can become a mentor on how to achieve hands from the heart.[7]

---

5. Senator Kennedy was shot on June 5, 1968, and died the next day, when the Sun had moved into the next degree.
6. Eduard von Hartmann died on June 5, 1906.
7. Jacopo Tintoretto died on May 31, 1594. Larger paintings have been done since by modern artists using methods that are much more rapid in execution.

C. G. Jung

Carl Gustav Jung died in this degree.[8] His mentoring can help us comprehend the true significance of the mythic structures of leadership, and how to learn from them. For example, Jung could well understand the magic cauldron as the collective unconscious, the great well of human possibility and destined future, a bottomless fount of creativity, horror, and love. He further understood that no event or person occurs out of relation with everything else. This viewpoint is a cure for all those who become enamored of technology and think that they can use that technology for their own pleasure privately without any impact on others. One could follow his example and build one's own house, using hand tools to fashion wood and stone, even carving stones for decorative sculptures. Jung built a house without electricity, wherein he meditated—making himself available for conversation with angels—and wrote his books. Jung would not have been afraid to call upon the dead as mentors for his researches, nor would he resist being called upon as a mentor—always respectfully, and with specific questions to which he can respond.

The mentors and spoilers mentioned in this chapter will be aroused by the Venus eclipse. Because the Venus eclipse occurs for the whole world, they are mentors or spoilers for the world. Each one has virtues that you can engage for dealing with the Taurus and Scorpio Images, and each one has spoiling qualities against which you can set clear boundaries. To Hartmann, you might say, "Yes to your brilliance and subtlety of understanding of the human condition, and no to your pessimism, as it turns everything dark gray." To

---

8.   Jung died on June 6, 1961, time unknown, with Sun at this degree of Taurus. If it was before six in the morning, the death Sun lay between 20 and 21 degrees, otherwise, quite close.

Jung, you might say, "Lead me on to a deeper understanding of my own regal nature—the truly greater king." Let us encounter Mozart more intimately.

Wolfgang Amadeus Mozart

The process of introducing you to a mentor to whom you are related by birth is described in much more detail in the Tea with Your Mentor package from StarWisdom.org. I would like to go through this process in a very compressed way with Mozart, as I consider his music a potential saving grace for the bumps that we may encounter from the Venus eclipse of the Sun.

When encountering a mentor, we can begin with a little study of his or her legacy. We can say about Mozart that he was a master of every form of music during his time, which is unique among composers. His legacy was illustrated by the stage play and film *Amadeus*, which, however, gave several untrue impressions, including the prominence of infantile behavior into adulthood. Mozart's last works were his greatest gift to humanity, especially *The Magic Flute* and his last symphonies, numbers 39, 40, and especially 41 (the "Jupiter"). He was working on the *Requiem* during his final illness; though this piece was finished by others, you can find much inspiration from Mozart there. These works dwell in the dark depths (from the themes of the Image in the Scorpion), then lift them up to the light. Don Campbell (*The Mozart Effect*) feels that listening to Mozart's music will cure every possible complaint, thus relating to the theme of the degree.

In our work with the imprints left by great men and women, we have found that the conditions of the last breath are important. They become a part of the person's legacy and must be dealt with specifically. Otherwise, they color the great gifts of the life too much. Mozart died in 1791, thirty-five years old, technically from "rheumatic inflammatory fever." He was more susceptible because of his work schedule, which alternated between hyperactivity, much envied by other composers, to melancholic depression, a pattern that today would be called bipolar disorder. Thus he also knew the challenges of the illnesses portrayed in the theme of the degree. He felt he was being poisoned. As he was often concerned with his health, and took many medicines, perhaps he had taken too much antimony, a metalloid element that was a popular drug in his day but is now known as a poison. At the end Mozart's body swelled with fluids to the extent that he could not move. He was bled by well-meaning physicians, losing up to three quarts of blood this way, which profoundly weakened him. He worked on the *Requiem* until the last hours, saying it would suit for his own funeral. He said, "I taste death," again perhaps a reference to the metallic taste of antimony. Cold compresses were applied to his burning head—he shuddered and went into a coma, dying two hours later. He was buried in a group grave, appropriate in his time, not in a pauper's grave as the film *Amadeus* shows. He died penniless.

While a Tea with Your Mentor study would go further, for the purpose of getting Mozart's general help around the time of the Venus eclipse of the Sun, we have enough to work with. We can acknowledge that Mozart is a brilliant guide and tutor on our path through the difficulties posed in the themes of the polarity of Scorpion to the Bull. He shows where we can go in the face of grave illness, dark streets full of the near-dead, and the relapse of the healed who then become ill again: one goes to the heart, and hands from the heart. First, actively say "no" to bleeding, to toxic medicines, and to dying young. Say "yes" to the power of bright and profound music to heal all ills.

Find Mozart's mentorship through his music. It can motivate you to learn how to extend hands from your heart. Can you learn to play the music of Mozart? Can you learn to sing Mozart? Can you learn at least to hum along with Papageno's songs from *The Magic Flute*, or the chorus of priests, or the aria of the Queen of the Night? You could learn the melodies from his *Jupiter* (Symphony 41), or some other piece. You can't really hum along to the *Requiem*, but you can certainly attend a live performance. Can you ask Mozart to assist in developing the clarity of thinking that balances with will, imbued with the mediator of healthy feeling? Can you ask him for an introduction to Venus, for surely he knew her intimately? Can you sing to the bleeding and the near-dead, extending the hands of your heart through the music of your voice?

7

THE SIGNATURE OF VENUS

"...the golden one, the Cyprian one...lover of laughter...she who subdues the race of mortal men, and the birds, and all the many animals that the land nourishes, and the sea nourishes."—HOMERIC HYMNS[1]

I n our examination of the Venus eclipse of the Sun, we have investigated the following alignment:

20–21 degrees of the Bull (Taurus)
—Sun—Venus—Earth—
20–21 degrees of Scorpio

We have concentrated so far on the quality of the energy, understood through an Image and a story, that lives in the polarity of 20–21 degrees of the Bull (Taurus) and 20–21 degrees of the Scorpion (Scorpio). Those themes are excited every time that the Sun lies on that polarity. The Venus eclipse empowers them more than usual. Thus we come to the question, "Who or what is Venus?" After we understand Venus, we can ask, "How does the presence of Venus accentuate the themes of the polarity of Taurus and Scorpio?"

---

1. From "Hymn to Aphrodite" (Greek name for Venus), *Homeric Hymns* (trans., Charles Boer, second edition, Dallas, TX: Spring, 1970), p. 69, the original being from Homer's time, approximately 2800 years ago or earlier.

Venus, Goddess of Love, of Beauty...

Mythology has collected a vast array of impressions from ancient times, a combination of actual experiences, intuitions, and make-believe. When we parade the stories and images of Venus, we have to resist being overwhelmed by these diverse reports, and take them in as something to be tested. In the later mythic systems, Venus and her Greek counterpart Aphrodite are differentiated from the many other expressions of femininity, including mature and divine expressions and mundane, primitive, and regressive expressions. Yet the early mythology recorded that the feminine was a unity—the maiden, mother, queen, and crone were different aspects of the same dynamic feminine divinity.[2]

The following reports of Venus are those that have made sense to me. I have read the many mythologies, compared them to my own intuitive experiences, and found a resonance in everything said here. Later in this chapter we will see how Venus becomes more prominent in conjunction with the Sun. For now, we deal with her reputation.

We hear that Venus comes closer to Earth than any other planet, and you can contemplate Venus in the sky as either morning star preceding the Sun or evening star, when she sets later than the Sun.

We hear that Venus loves copper, a warming and intimate metal; when you look into a mirror of burnished copper, you have a very different sense of yourself than you get from the cooler nickel surfaces presently used. Indeed, Venus rules mirrors. I recently visited a hotel in New York City that had remodeled its rooms but could not rebuild its antiquated elevators, which were small and slow. A housekeeper explained that they had put up mirrors at all the elevator doors, and the complaints about the long waits diminished significantly. Venus ensnares you in appearances, and dissolves time.

---

2. Jane Ellen Harrison, *Prolegomena to the Study of Greek Religion* (Princeton, NJ: Princeton University Press, 1991; original third edition, 1922), chapter 6.

Narcissus gazing at his reflection (from chapters 2 and 3) has lost all sense of time.

Venus rules lovely appearances on the outside, as well as the loveliness that lives on the inside.

Venus loves laughter. Fridays were named after the Norse version of Venus, Freya. Venus has gifted human speech with the first vowel, "ah," which she continues to hold and emanate—soft arms open to another, preparing for an embrace. "The sound of 'ah' is white, it flies openly; multiple forms of opening up of the arms express it; in it is a fullness of the soul; reverence, worship, wonder; the apprehensible beginning is 'ah'; everything else is lower."[3]

In ancient Greece, Aphrodite, and the same goddess later in Rome, Venus, was known as the origin of beauty, love, grace, and relationship. Beauty! Love! Sensuality! Grace! When one finds the source of these broad secrets of existence, one can only feel devotion. That is what is appropriate to Venus, and what she demands—devotion, a healthy reverence for the wonders of creation. Eros, often depicted as a young male, or in its form as Cupid, a young male child, always acts as the agent of Venus, and thus is really the same as Venus. Her realm is everything erotic, everything life-creating.

The keyword book of astrology has over 1,100 traits and objects that have been associated with Venus and that stretch our understanding of her realm, including smiles, precious stones, pacifism, gaiety, negligees, songbirds, analogies, and artists.[4] Hers are all trees, especially the apple tree, and the apple isle, Avalon; hers all the animals. She is the lady of the beasts. She is the white goddess.[5] Venus is a messenger of the Sun—she focuses the Sun's

---

3.   Andrej Belyj, *Glossolalie: A Poem about Sound* (Dornach, Switzerland: Rudolf Steiner Verlag, 2003), p. 205.

4.   Rex E. Bills, *The Rulership Book* (Tempe, AZ: American Federation of Astrologers, 1971), pp. 198–211.

5.   Robert Graves, *The White Goddess* (NY: Farrar Straus and Giroux, 1948, amended and enlarged). Buffie Johnson, *Lady of the Beasts* (San Francisco: Harper, 1981).

life-giving warmth and fecundity. Powers of germination, flowering, reproduction, and also of dying into the earth to be born again—all these belong to Venus.[6] She gives us fruits and flowers, and everything juicy, perfumed, and colorful. As Lucretius wrote to Venus, "Only through you are living things conceived and because of you they rise up to bask in the light of the Sun."[7] The goddess has powers to conceive you, to lift you up into the light, even to open the whole realm of heaven to your gaze and your participation:

> Ishtar, the Goddess of Morning, am I;
> Ishtar, the Goddess of Evening, am I;
> (I am) Ishtar, to open the lock of heaven belongs to my supremacy.[8]

All the aspects of romance—the buzzing of the bees in flowers, the courtship dances of colored birds, and the heart arrhythmias of adolescents—are incited and overseen by Venus. Lucretius again: "Under your spell, all creatures follow your bidding, captive, eager even."[9] Does this describe your personal experience? When you are captive, and do another's bidding, it means that your will is under another's control. Venus rules through the feeling realm, yet her effects are often seen in the realm of will.

Venus governs money and trade. Why? Because all exchange involves a social relationship. Though Pluto governs excessive riches, large stashes of gold, and kleptocrats (those who steal in order to rule), Venus governs the

---

6. Venus was identified with Quetzalcoatl in the mythos of Central America: The Archive for Research in Archetypal Symbolism, *The Book of Symbols* (Cologne: Taschen, 2010), p. 688.

7. Titus Lucretius Carus, *De Rerum Natura (On the Nature of Everything)* (trans. David Slavitt, Berkeley: University of California Press, 2008), Book I, originally written around 50 BCE.

8. From an ancient Babylonian hymn, in Harold Bayley, *The Lost Language of Symbolism* (New York: Citadel, 2006, original 1912), 177. Venus is the morning star and evening star at different times of year, and here Ishtar is found to be another name for Venus.

9. Lucretius, *De Rerum Natura*, op. cit., Book I.

simple exchanges between people. When you buy something from some-one who made that thing, or when you haggle over a price in a third-world bazaar, you engage the other in the realm of feeling. Westerners used to pay-ing the price listed on the tag have to learn to bargain, which they will soon discover means opening up to a wide range of feelings in interaction with the seller, who feels incomplete without this interpersonal struggle.

Venus has to have an exchange of energy, back and forth. Voyeurs only take energy without reciprocating, and are universally condemned. Even spectators at a sporting event participate by screaming and singing.

VENUS IN THE CINDERS

To accomplish these realizations of beauty in the light may require time among the cinders, as with Cinderella, where one goes down into the depths of human experience. Cinderella is truly Venus unrecognized, who has forgotten herself and must be reminded of her divine nature.[10] We travel with Venus into these depths, where taste goes foul and vision becomes all grays...where sexuality becomes addicting, and we take on other addic-tions to one or another of Venus' sensations. Then we can rise up into the light, all the more wonderful because of where we have been. Ishtar/Venus descended into the land of darkness where "they behold not the light, but dust is their bread and mud their food."[11]

The powers of Venus to evoke sexual fantasy, display, and behavior sometimes lure people into the dark part of her world. I have found that many people are afraid to speak of this part of Venus, preferring the serene beauties of a radiant but floating goddess. But the underworldly

---

10. Bayley, *Lost Language of Symbolism*, op. cit., chapter 8, follows the tale of Cinderella in its many variants (and there are hundreds) from every culture, finding there the descent of Venus/Ishtar/Inanna.
11. Bayley, *Lost Language of Symbolism*, op. cit., p. 176. In chapter 4, we saw that Inanna did the same.

Venus—in her guise as Persephone, consort of the demon of the dark world—holds great power. We shall touch upon this polarity much more deeply. Pornography is the largest category in the world of the Internet (Jamshid's cauldron). Many of those sites are "down and dirty," that is, in the cinders. To save themselves from Venus' enchantments, some people insist on covering all women from head to toe or censoring every image and word to remove references to sex. This doesn't work, and it ought not work, as the foundation of sex is necessary to hold everything together. The picture of pubescent girls gathering flowers in a meadow, their hair flowing freely in the sunlight as they run from place to place, singing of flowers and sweet things, brightens every human soul. At the dark extreme, in the dark of secret places hidden by night, sex workers function as "sacred prostitutes," playing the important role that they have performed through the ages in re-civilizing warriors once they have returned from battle.[12] Sex lures even the most hardened warrior, including the men who have fought, been wounded, and harmed others in the world of commerce. These extremes and everything in between play out every day in every place on Earth. When one sees pornographic images from the past, such as are on display at New York City's Museum of Sex, one can only be astounded that nothing has changed over millennia. The powers of Venus extend beyond sacred prostitutes, into areas of strange perversions that ensnare body and soul, a theme to which we will return as we move up and down the scale of her expressions, beginning with the positive powers that Venus exerts in our lives.

VENUS AS SOPHIA AT HOME IN FEELING

To those who know her deeply, Venus represents or is the same as Sophia, the heavenly principle of divine wisdom, the stuff of all creation, the core

12. Nancy Qualls-Corbett, *The Sacred Prostitute* (Toronto: Inner City, 1988).

power of life.[13] Ralph Waldo Emerson summarized this in his essay on love: "Every soul is a celestial Venus to every other soul."[14] Read that line aloud, as Emerson did repeatedly in his lecture tours, and feel that Venus in her best role unites through relationship one with another in the world of soul.

If we think of the human being as divided into powers of thinking (mostly in the head), feeling (at home in the torso), and willing (expressed through the limbs), then Venus is most at home in the middle, in feeling and emotion, in drama and music.

We can further divide the feeling world of music into a thinking/feeling component in melody, a feeling/feeling component in harmony, and a willing/feeling component in the beat. Venus continues most potent in the center, in harmony, the sensual relationship between two or more notes with each other—and two or more people with one another, finding a harmony between souls.

Venus shares the middle realm with the Sun. An eclipse of Venus before the Sun sounds two widely spaced octaves of the same powerful note.

To Venus one prayed for power of enchantment, for the power to banish thinking and induce an ache in the breast, a desperate longing: "Give me the kind of song that seduces, please," pleads the ancient bard.[15] To Venus one prayed for a mate. From Venus one received advice about how to get one, including magical spells and potions—to increase one's attractiveness, to enhance the skills of sex, to ensnare the right mate, and then to keep that mate.

Marketers seek the help of Venus to sell their wares. You can enter any auto repair shop and find calendars posted on the walls featuring metal car parts with women in bathing suits. Several times I have heard mail-order

---

13. From the *Pistis Sophia: A Gnostic Gospel* (trans., G. R. S. Mead, Blauvelt, NY: Spiritual Science Library, 1984), chapter 137, p. 298.
14. Ralph Waldo Emerson, *Emerson's Essays*, op. cit., 121, the opening line of his lecture "Love."
15. The third "Hymn to Aphrodite," *Homeric Hymns*, op. cit., 83.

experts speak about garden tools (a business in which I was once involved) in terms of how "sexy" they were or could be made to seem. Different consultants at different times counseled this same approach, beginning with the advertisers' formula, "It's not the steak that you sell, you sell the sizzle." When I asked, "Why?" they answered, "Because it's more sexy!" (They don't say "sexier," because "sexy" is their measure, whether or not something has got "sexy.")

Thus Venus deals, more than the other gods, with *erotic phantasms.*

Erotic Phantasms

This section introduces some difficult concepts, yet gives us a deeper understanding of Venus' power in the life of the psyche.

In a close reading of Marsilio Ficino and especially of Giordano Bruno, the Romanian philosopher Ioan Couliano realized that these Renaissance philosophers understood the role of phantasms in culture, what Blavatasky, Alice Bailey, and Rudolf Steiner would later call thought-forms and concepts.[16] The Avatar technique has for twenty-five years trained people to identify thought-forms as having size and shape, that is, boundaries, as well as texture, quality of movement, etc.[17] Once you can experience them as things, you can deal with them. Usually these phantasms lie unseen within our mental and emotional worlds, organizing all of our experience yet transparent. We see through them to what they are about. In the sixteenth century, Bruno explained clearly that the foundation for phantasms was Eros—life force—the power of Love, of Venus, applied to matter. Eros pours into

---

16. For example, Rudolf Steiner in "The Mission of Spiritual Science and of its Building at Dornach," lecture of January 11, 1916. Ioan Couliano, *Eros and Magic in the Renaissance* (Chicago: University of Chicago, 1987).

17. www.Avatarepc.com, which means "Avatar—Enlightened Planetary Civilization," an excellent training in how to identify and master erotic phantasms, though they don't call them that.

us and spills over in abundance to affect our thinking, feeling, and willing. Erotic phantasms include not only our ideas about beauty, pornography, and overt sexuality, but every kind of desire, such as adulation of celebrity, propaganda by governments and salespeople, corporate branding, and patriotism. On a personal level, Eros enters thought-forms in craving for chocolate, rage about a wrong, fetishes of all sorts, scenes from a movie repeating over and over in one's mind, compulsion to clean, yearning for comfort, addictions of all sorts, stories about others (gossip, grudges, judgments, also crushes and infatuations) and stories about oneself (justifications), attachment to one's identity and appearance leading to cosmetic surgery and indeed all cosmetics, "my favorite things" (from *The Sound of Music*), and many other patterns of desire and attachment. Bruno instructed how the magician can manipulate others through the power of erotic phantasms, and his text reads like a primer for politicians and merchants of our day, four hundred years later. Bruno's work "deserves to have the real and unique place of honor among theories of manipulation of the masses."[18]

In our longing (Eros) for spirit, we pour out our selfhood into our sensory perceptions, seeking satisfaction in matter. We make our perceptions into concepts, which then govern how we integrate new perceptions, building up the concepts into dense forms. People think of their concepts—or, in Bruno's terms, phantasms—as what they picture, but there are two aspects to a concept, the picture and the motive behind it. For example, if I have a yearning (Eros) for an ice cream cone, I can picture the thing before me. There is the ice cream cone in my imagination, and I then begin to plan how I will get one. I may think that the concept is the same as my picture of an ice cream cone, because that's what comes up whenever I think of it. I feel the desire or craving for what's pictured. But the picture is only the facade of a motive force—a movement of Eros within my being—that I do not see.

---

18. Couliano, op. cit., p. 89, about Bruno's *De vinculis in genere* (of chains in general).

Every picture (phantasm) of things in the world has such a dynamism of Eros moving behind it.

As an example, an advertising jingle in itself is silly, but it stays in the mind because a magician (the advertiser) has understood how the jingle will engage Eros working in us. Eros is the fuel, and the phantasm the vehicle.[19] We don't see the fuel; we see only the vehicle.

Beauty itself has become an erotic phantasm. When I ask Google Images simply for "beauty," I get hundreds of photographs of the faces of young women and nothing else. None of these images is real, not only because of physical cosmetics, but because they have all been scrutinized by computer and altered to homogenize the skin, take away wrinkles, add sparkle to the eyes, on and on.

19. The Freudians have studied this dynamic from the point of view of libido (Eros, vital force) moving through the ego (structure of the personality) to cathect (attach) to objects, imagining it all as a unique expression of the individual. In his four-part series for the BBC, "The Century of the Self" (2002), Adam Curtis attaches responsibility for modern marketing of products and politicians to a careful application of these ideas.

The vehicle—the phantasm of the face of beauty—papers over the erotic dynamism working beneath. "Under your spell, all creatures follow your bidding, captive, eager even."[20] Once you see how the spell orients you to outer appearances, you can choose to encounter the life force directly, the Eros beneath. You can perceive beauty not in the many images of faces, but through them. And then you understand that pretty faces limit the beauty of Venus. As the only feminine deity expressing through planetary movements, Venus is not only the young woman, the sexually energetic perfection that we see in these images. Venus is all women. She is the little girl in the pink dress; she is the maiden collecting flowers for a garland; she is the young seductress—you yearn for her delicate fingers to caress your face; she is the mother growing another human being in her firm womb and nourishing the baby from her soft breasts; she is the upright queen in her power; she is the wrinkled crone in her wisdom. She is all of these. The outward beauty of her early years moves more and more to the inside. You can see the underlying Eros from Botticelli's Venus—or from those hundreds of faces from Google or the dozens of magazines engaging you at any newsstand—in every age of woman. In fact, to discover Venus, you must make this shift, to see the Eros power behind the exterior picture to see what lives inside, without picture.

Too often people who perceive the full extent of what Bruno would call Venus' chains recoil completely, and reject everything lovely, warm, and juicy. The Calvinists did this, as well as many other groups historical and modern. In their fear of the excesses of Venus and Eros, they have rejected the generous wellsprings of life. They have become dry, cynical, intellectual, infatuated with antisocial technologies, and prone to phantasms of ideology that justify their stance. You actually can't escape phantasms—this is how we think. The only solution is to embrace them, and learn to distinguish

---

20. Lucretius, op. cit., Book I.

between the appearance—the ice cream cone or the ideology—and the Eros living behind it.[21]

WHAT OF THE MEN?

And what of the men? Are they left out of creation's great display of diverse liveliness? Men must understand that their bodies are derived from Sophia, both in substance and form. Their bodies belong to the divine feminine, and are borrowed for the particular expression we call masculinity. Participation in the wonders of creation does not mean that males must become effeminate or, worse, "nice men." Or that they cultivate their "feminine sides." Males must give gratitude where it is deserved, that is, to creation, and enjoy their particular expression of the primal life force that exceeds the limitations of gender. The worldly feminine of the marketplace has taken ownership of the divine feminine, over which it has no exclusive rights, for at the level of Venus in her highest expression, her domain is resource and refuge for all earthly genders. Let the women express the divine feminine in their mundane femininity, and the men in their mundane masculinity, with neither ever forgetting to whom they owe the fact of their existence.

A young woman once confided in me that she worked as an exotic dancer in a strip club. I asked her how that was. Immediately she blurted out, "Men are pigs!" Let's understand the scene: We have music (sphere of Venus), naked women (Venus), for a while dressed in lingerie (Venus), moving seductively (Venus), being paid money for their undulations (Venus), in a soundscape of boom-boom-boom beat (emphasis on the realm of the will, and obliteration of thinking), that is loud (emphasis on the will, diminished thinking), in the dark (emphasis on the will, diminished thinking), with a rule of don't-touch-or-a-bouncer-will-hurt-you (also emphasizing the will), with alcohol (diminished

---

21. Max Weber's essay "The Protestant Ethic and the Spirit of Capitalism" (1904) makes the connection between the cooler mentality of the less erotic and anti-erotic Protestant faiths and the rise of materialistic capitalism.

thinking, impulse rising). The stripper's comment recalled to my mind the encounter of Odysseus' men with Circe the enchantress. She used the powers of Venus to turn the men into pigs, beings completely ruled by their appetites, all will to the exclusion of thinking and feeling. How does this happen that men become pigs? Who initiates that behavior? The men and women in this scene both participate in the erotic phantasm, and it differs only a little from the parties that occur in every culture and country of the world.

I thought also of the tales of mermaids and sirens luring men to their deaths through their seductive singing and naked bodies, perched on dangerous rocks in the sea. The men are unable to exercise their will with the women. They can only be voyeurs yearning to touch and possess, enchanted by the illusion of Venus.

VENUS AND ILLUSION

People seldom take steps toward the true Venus, to the divine feminine. They rather become entangled in the outer display. Thus to the philosophers Venus became known as the source of illusion.[22] The *Oxford Classical Dictionary* relates Venus' close connection with wine, the physical tonic of illusion.[23] But Giordano Bruno goes deeper in his explanations, delving to the dynamics of love itself: "The strongest chain is that of Venus.... [A]ll bonds relate to the bond of love, either because they depend on it or because they are reduced to it.... The chain of chains is love."[24] We can all relate to that warning,

---

22. For Aphrodite as the origin of illusion, see the classical scholar Peter Kingsley, "The Spiritual Tradition at the Roots of Western Civilization," *Anti-Matters*, 1(2), 2007, pp. 111–141, and *Reality* (Point Reyes, CA: Golden Sufi Center, 2004).

23. *The Rulership Book* by Bills, op. cit., p. 158, agrees that wine and wine merchants are linked with Venus, though especially sweet and light wines.

24. Cited in Ioan Couliano, *Eros and Magic in the Renaissance* (University of Chicago, 1987), 97. It seems that Bruno never came to know true Love. "I loathe the vulgar herd and in the multitude I find no joy," *On the Infinite Universe and Worlds*, the introduction.

recalling how attraction, passion, sex, and the compulsions of love have made fools of all of us.[25] This is the key to the quote from Homer's hymn at the beginning of this chapter: Venus "subdues" by exciting all creatures with the fantasies of love. They become "captive, eager even"—eager for their own captivity in the passions of Venus.

Rudolf Steiner equated Venus with the Grand Illusionist Lucifer, the "light-bearer" who excites us with insights and visions, yet also wishes to lift us up and away into rosy mists of fantasy, thus distracting us from our true tasks on Earth.[26]

Thus Venus' tree, the apple, became identified with the Tree of Knowledge of Good and Evil, which, with the help of a serpent—or because Venus herself had the qualities of both serpent and apple—seduced Eve to take and eat. The apple, that is, Venus' essence, was blamed simultaneously with dissolving the illusion of Paradise through knowledge of good and evil, and with creating a new illusion of blood, sweat, and tears.[27]

Yet we know that passion, once understood and worked with positively, can create the best in us. If we can develop a strong self—a personal sense of standing firmly in our unity—then we can have a mature relationship with another, the Two. From this foundation we can enter into a communion with Nature and spiritual worlds, a larger ONE. Thus, beginning at the foundation with the one person knowing himself or herself:

---

25. The *Oxford Classical Dictionary* (p. 1587) connects the name Venus with *venia* (gracefulness), *venerari* (to exercise a persuasive charm), and *venenum* (poison). This last relates to the way in which people are poisoned—by drinking or eating from one whom they trust, that is, whom they love. *The Rulership Book* by Bills, op. cit., does not link poison with Venus, but rather with Neptune.

26. Rudolf Steiner (*Man in the Light of Occultism, Theosophy, and Philosophy*, p. 65, 168) speaks of Lucifer as the same as Venus, connecting the seductive feminine with the master illusionist. Christ has also been likened to the morning star, Venus, when she rises before the Sun in the east (2 Peter 1:19, Rev. 22:16, Rev. 2:28).

27. Robert Graves, *The White Goddess*, op. cit., p. 253 and passim.

Begin with: One—foundation in individuality, 1 = 1
Progress to: Two—healthy relationship, 1 + 1 = 2
Realize: ONE—communion, 1 + 1 = 1

We can flow by choice up and down this ladder. This is the recommendation of my book, with my wife, *One Two ONE: A Guidebook to Conscious Partnerships, Weddings, and Rededication Ceremonies*. It seeks to help others to create healthy and active relationships, wherein one improves oneself through the other's reflection, making true communion possible.

We can cultivate devotion—and must do so, to save our heart from drying up in this too-fast world. This can look like illusion. The "just the facts, ma'am" types, the hard-headed ones, picture devotion as ignoring tough reality, as fanciful, as weak. Yet they must observe the miracles around them—the birth of a child of any species, the depth of interrelations of life forms in a handful of garden soil or a handful of galaxies—and feel them as miracles. To save one's heart, to save one's soul, one must encounter the "aboriginal abyss of radical amazement,"[28] and from this comes devotion. Yes, devotion can appear as illusion, and, yes, we must simply make the effort to keep them separate. We must pass through Venus' enticements to unite with that light, for, as Lucretius wrote over two thousand years ago, "nothing comes forth to the light except by you, and nothing joyful or lovely is made."[29]

Thus Venus brings a mixture of wisdom and lust, of ecstasy of the body and ecstasy of revelation of truth. Venus encourages gnosis, that is, personal knowing of divinity once Venus has opened the lock and the door to heaven. The most convincing philosophers—which means lovers, *philia*, of wisdom,

---

28. Abraham Joshua Heschel's wonderful phrase from *Man Is Not Alone* (New York: Farrar, Straus, and Giroux, 1976), p. 13. Each word in this phrase deserves meditation and action.
29. Lucretius, op. cit., Book I.

Sophia—confess that first they are lovers. Their erotic phantasm can be trusted, for they see the Eros behind all concepts and behind all disguise.

VENUS AND THE WILL

Venus as goddess is known for her golden apples as vessels of sweetness. Into a gathering of gods and goddesses, the goddess Eris threw a golden apple inscribed, "To the most beautiful."[30] The goddesses wrangled, all aspects of the same goddess though dividing in argument about who was the most beautiful. They nominated young Paris to decide who was most beautiful. Bribed by Venus' promise that she would deliver to him the beautiful Helen, Paris awarded her the apple. She planned how Paris, a prince of Troy, would abduct Helen, then wife of the Greek lord Menelaus. Thus began the Trojan War, as the Greeks united to get Helen back. We see in this story how little Venus cares for politics or familial ties. When you are Venus-struck, you act from the will and not from thought. You are enchanted by Venus, able only to confirm that Venus is the most beautiful.

Rudolf Steiner cautioned that we not take Venus' illusions, that is, Lucifer, into our will—our deeds and actions.[31] This warning relates to the Image at the Bull, the picture of Jamshid and his magical tools performing magical deeds through his will in service of the improvement of humanity. The warning relates to everyone who acts on an ideology, or who loses himself or herself in attractive cyber-worlds.

For humanity, it's too late to heed Steiner's advice. We've already taken illusion into our will in every possible aspect of our culture. However, we can heed the warning and begin to withdraw, to balance this onrush into

---

30. We have to consider that Eris is an aspect of Venus, as this aroused the typical Venus dynamics of jealousy, competition, and the prototypical beauty contest.
31. In Steiner's "The Guardian of the Threshold" (*Four Mystery Dramas*, scene 10, 123), Maria says: "A man may turn toward Lucifer and warmly feel inspired by his bright glory ...but he should not take Lucifer into his will."

unconscious will, to come to the deeper offering from Venus of true inner beauty. In the Book of Revelation, a great gift is offered from divinity to humanity: "I give him the morning star," Venus shining brightly before the sunrise, Venus as the harbinger of the beauties of the day, clear-headed, full-bodied enjoyment of life.[32] Venus' gift of beauty, generative power, and relationship ought not be missed by addiction to her surface attractions.

THE MANY POWERS OF VENUS

These many features of Venus' influence can be arranged in a kind of continuum, from the darkest and most enchained at the beginning, then through various versions of orientation to the feeling heart, and ending with a retreat into thoughts, the thought of beauty:[33]

> Sex addiction, compulsion, entrapment and the fear that it arouses; obsessions of various sorts, including with one's appearance, thus eating disorders; pornography, lust, enthrallment; other addictions; not a trace of thinking except repetition compulsion; the dark Venus as pictured in chapter 1;
>
> Coquettishness, plunging necklines and rising skirts, teasing and titillation; the bump and grind of rock music concerts; cosmetics and the other common phantasms mentioned above which, though common, are nonetheless a part of Venus' darker enchantments;
>
> Balanced, healthy sexuality, fecundity, juicy generativity; awareness of the mythic powers of life and the urge to create more life, thus also creativity;
>
> Participatory appreciation of beauty in the sensory world; sensuality; those who garden with their hands; physical and emotional intimacy with friends and others in one's karmic group;

---

32. Rev. 2:28; also in 2 Peter 1:19 and Rev. 22:16.
33. In the language of Tresemer (*Star Wisdom & Rudolf Steiner*, op. cit.), the dense beginning would be within the realm of the Hardener, and the floating fantasy at the end in the realm of the Illusionist. In anthroposophic language, the Hardener is Ahriman and the Illusionist Lucifer.

Capacity to feel the beauty in all and everything;
Radiance as a pure messenger of the Sun, uniting soul to soul; the bright
    Venus as pictured on the book cover;
Pure inspiration from the magical fountain of Life;
Floating reverie in fantasy of pleasure, ecstasy; not a trace of will forces,
    entirely consumed in a narcissistic illusion.

These are all Venus in action, Venus as the divine feminine, as Sophia and as Shekinah. Each must be understood and mastered in order to find the way through the phantasm to the life force (Eros) behind and beneath.

I have been most impressed by the training for women called The Path of the Ceremonial Arts, at the StarHouse, in the way it works all the different experiences of the divine feminine.[34] That training opens the student to the fearsome and the sublime, and thus teaches the capacity to know oneself in freedom in relation to these potent life forces. Only then can one perform true ceremony in the realm of the heart and beauty, when imbued with devotion to the wellsprings of creation.

From Reputation to World Events

So far we have shared Venus' reputation. The ancients knew the archetype and ideal of the goddess, and found her rhythms in those of the planet that we call Venus. How do we observe Venus' activity? How can we find the signatures of Venus in actual events and phenomena?

Venus shows herself most powerfully when in relation to the Sun. As Sappho once remarked, identifying with Venus: "I love luxuriance—this, and lust for the Sun has won me brightness and beauty."[35]

---

34. See http://www.pathoftheceremonialarts.org.
35. Sappho fragment quoted by Gregory Nagy, *Greek Mythology and Poetics* (Ithaca, NY: Cornell, 1990), 261. Sappho, who lived in the sixth century BCE, would have had the name Aphrodite in mind, whom we see as the same as Venus.

Taking a hint from Rudolf Steiner that every action performed by Christ Jesus was in complete accord with the astrological configurations of the heavens, I extensively researched the life of Christ Jesus, particularly the last three and a half years, from the baptism in the River Jordan to the Mystery of Golgotha. There was no eclipse of the Sun by Venus during the lifetime of Christ Jesus. However, we may correlate the event in 2012 to the times during Jesus' life that Venus did conjunct the Sun, both inferior—Venus between the Sun and Earth, closest to the phenomenon of the eclipse—and superior—Venus behind the Sun from the Earth's vantage. During the inferior conjunction, Venus appears to move very swiftly in a retrograde fashion, that is, backward through the progression of the zodiac, about one day within one degree of exact conjunction.[36] During a superior conjunction, Venus travels with the Sun from the Earth's point of view, and can seem to be in a one-degree orb for many days—in one instance cited here, nine days.

> A superior conjunction looks like this: Venus—Sun—Earth.
> An inferior conjunction looks like this: Sun—Venus—Earth.

Rudolf Steiner suggested that Venus in inferior conjunction stimulates the head-pole, that is, the nervous system and the organization of thinking, concepts, and illusions. In contrast, the superior conjunction, with Venus

---

36. When a planet moves retrograde, it doesn't actually change direction. The Earth, moving swiftly along its path, makes the other appear to move backward through the zodiac. I explain these things more clearly in the StarWisdom 101 course from www.StarWisdom.org. This change of point of view, when another body seems to move backward, has been identified as a stimulator of the issues excited by that planet. Thus Mercury-retrograde is often identified with difficulties with communication. I view retrograde motion as accentuating the qualities of a planet's influence and making them felt more strongly, rather than undermining them.

further away, on the other side of the Sun from the Earth, stimulates the limbs, that is, the will-pole, the initiator of deeds.[37]

The Venus eclipse is a very specific form of the inferior conjunction where Venus glides not only between Sun and Earth, but exactly in the Sun's path, neither above nor below the Sun but across the face of the Sun. Thus we might expect that the 2012 eclipse will have more to do with our thinking, concepts, and illusions.

During Christ Jesus' lifetime, Venus crossed three times between the Sun and the Earth, that is, in inferior conjunction (most like the eclipse that will occur on June 6, 2012), and three times in superior conjunction. I have arranged these into short vignettes, followed by commentaries, as we have done with other planets.[38] We will build up an understanding of Venus as we go through these vignettes, organized by theme.

As the research on Venus unfolded, I realized that another Venus position proved important. The conjunction of Venus with the Sun was supplemented in many themes by Venus conjunct the Eye of the Bull, Aldebaran, the Life Star, at 15 Taurus; also Venus conjunct Aldebaran's opposite, Antares at 15 Scorpio, the Star of Death and Rebirth, the Heart of the Scorpion/Snake/Eagle—for Scorpio was known as all of these. These two main Royal Stars of Persia lie exactly opposite each other, and form the basis for the divisions of the zodiac.[39]

---

37. From a close reading of Rudolf Steiner, *Man in the Light of Occultism, Theosophy, and Philosophy* (Blauvelt NY: Garber, 1989), pp. 165–167. The realm of feeling is stimulated especially when Venus lies at its furthest from the Sun from the Earth's point of view, a location that we will study in the future.

38. The methodology of this approach is discussed in "Signature of Saturn..." and "Signature of Jupiter..." in previous issues of *Journal for Star Wisdom* (and its predecessor, *Christian Star Calendar*). The methodology is also available at www.StarWisdom.org. More about the spiritual side of Venus can be found in Lacquanna Paul and Robert Powell, *Cosmic Dances of the Planets* (San Rafael: Sophia Foundation Press, 2007), pp. 133–161.

39. On the use of the Aldebaran–Antares axis to set out the zodiac in ancient times,

Now we move from the images, character traits, and the reputation of Venus, to actual events when she was prominent, by virtue of the amplification by the Sun. I have organized these by the kinds of themes that arose when Venus was prominent.

1. THEME OF DESTINY, DESTINY GROUPS, THE DIVINE PLAN, AND LISTENING

1A: Imagination: Inferior Conjunction in the Goat: Listening to the Past and to the Future

A month previously the teacher had come down from his time alone in the desert mountains, from his encounter with demons and with the Father God. He did not speak about this to anyone, but his mother knew of his retreat in the wilderness. She could feel his movements, both outer and inner, where he traveled and what he felt. She tried to understand what he intended, where it was all going. At the great wedding where friends and relatives were gathered at Cana, he had done something magical. She had helped, by instructing the servants to heed his commands. He had challenged her, saying, "Woman, what relationship is there between you and me?" She remembered this question, and had wondered about it, because

---

see Robert Powell, *History of the Zodiac*, op. cit. On the qualities of Aldebaran and Antares, see David Tresemer with Robert Schiappacasse, *Star Wisdom & Rudolf Steiner*, op. cit., early chapters. Roberto Calasso, *The Marriage of Cadmus and Harmony* (New York: Knopf, 1993), p. 206, has this to say about the polarity of Taurus and Scorpio: "The cosmos pulses back and forth between snake and bull. An enormously long time would pass before the snake, Time-Without-Age, was followed by the drumming of the bull, who was Zeus. Then a much shorter time before the bull Zeus coupled with Demeter to generate a woman in whom the nature of the snake pulsed once again, Persephone. And hardly any time at all, the time it takes for desire to flare, before Zeus, realizing that the baby Persephone had become a girl, transformed himself into a snake, coupled with his daughter, and generated Zagreus, the bull, the first Dionysus." All these couplings were overseen, of course, by Venus (as Aphrodite), mistress of procreation, even that between the gods.

something extraordinary had happened, all agreed on that, though none could agree on what had actually happened. Who was this man, her son?

On this day, after traveling with his students, the teacher came to meet with his mother, and spoke with her long into the evening. He told her what he had experienced while in the mountains, phenomena inner and outer, birds and stars overhead, the courageous plants clinging to the moist cracks beneath his feet, all speaking to him. He spoke of what he was going to experience in the coming few years, which would come to a close in a most bloody way. He spoke of events in the world of men and women and also within his own soul.

A few months previously, he had shared his feeling that the spiritual forces that had been such strong supports in the past had abandoned humanity. The ubiquitous music of divine support could no longer be heard. What remained behind were empty rituals. The Essene community had succeeded in keeping out demons and distractions, who then attacked the rest of humanity all the more strongly. He suffered immensely from this realization of what humanity had become. This motivated him to seek his own destiny. He mixed his words with sounds unknown to Mary, sounds that she could not identify though she understood their meaning. In this conversation, Mary had felt herself expand in light, in communion with the Mother of the world.

This evening he spoke of the future, his future and the future of humanity. It was an immense story, containing horror and bliss, ending with a sword piercing Mary's heart so that it would open more fully to others. The vast tapestry of the story linked the depths of the earth with the heights of the heavens, all interweaving with human actions and words. She knew that she would not be able to speak to others about what he had told her this night. She knew that she would knit together the followers that her son attracted, the poor and the rich, the vital and the limping, the stragglers and the scholars, the men and the women, into a strong community with her strength at its core. And then, after everything he told her had unfolded, and the followers had all fled in terror, she would pull them together again, and rebuild.

Comment: Prior to this meeting, an intimate sharing occurred just before Jesus decided to seek the water initiation offered by John the Baptist. At that meeting Venus was conjunct Antares, the Star of Death and Rebirth, thus in our terms akin to Venus before the Sun—accentuation of both influences.[40] With his mother, Jesus used unusual words and sounds that they both understood.[41] This conversation between mother and son—technically, the Solomon Mary and her stepson[42]—was admired by Rudolf Steiner as the intimate sharing between the Sun God—the Christ light—and the bearer of the divine feminine, Sophia.[43] Listening, the meeting of soul to soul, the discovery of karmic groups to which one is related, the connection with Sophia—these are all present in this series of talks, beginning with this first meeting.

The second intimate sharing, the one imagined in the story above, occurred in the short span of time when Venus was in inferior conjunction to the Sun, just five weeks after the forty days in the wilderness, and ten days after the turning of water into something precious at the wedding at Cana.[44] Though Jesus gave little hints about his expected trials to his disci-

40. Venus 15 Scorpio, Sun at 0 Libra 50, thus also a maximum distance from the Sun (from the Earth's point of view), putting this event also in the heart realm.

41. Peter Selg, *Rudolf Steiner and the Fifth Gospel: Insights into a New Understanding of the Christ Mystery* (Great Barrington, MA: SteinerBooks, 2010), p. 142.

42. The story of the two Jesus children and the two mothers Mary can be found in Robert McDermott's introduction to Rudolf Steiner, *According to Luke*; Powell's *Chronicle of the Living Christ*; and my *Star Wisdom & Rudolf Steiner*, op. cit.

43. Selg, op. cit., pp. 56, 73, 74, 145.

44. This sharing occurred on January 8 CE 30, Sun and Venus at 20 Capricorn 36. One degree either way, that is, from 19 Capricorn 36 to 21 Capricorn 36, goes from noon on January 8 to 5:40 p.m., January 9, not quite 30 hours.

ples and audiences on many occasions, the telling of these tales to the bearer of Sophia/wisdom herself was the most comprehensive. The light-unfolding spoke; Venus listened. In this fragment, we see Mary's central importance in fulfilling the life mission of Christ Jesus.[45]

The planet and sphere of Venus creates the deep inner spaces of the soul to embrace all that is beloved. A summary of the past was given in the first meeting, when Venus lay conjunct the distant powerful star Antares. When Venus lay conjunct the Sun, the sharing became more immediate and oriented to the future. Prophecy poured forth, not fearful predictions, but rather ones that engaged the will to look courageously at what would unfold. The key to this kind of prophecy is the quality of listening, soul opening to soul. Then both can participate in destiny unfolding, sharing the sense of responsibility that prophecy can bring. When prophecy flows, one becomes larger than the usual: divine Christ-light meets divine wisdom, Sophia.[46]

Venus in the head-pole stimulates pictures of the future that ask one, "Will you engage in this destiny for the good of all?" Here we have the best example of an important trait of Venus: "Venus listens,"[47] which we could amend and explain, "In the realm of Venus, Venus listens, Venus speaks, Venus listens." If we understand the different natures of Venus interconnected—the planet, Venus as Aphrodite, Venus as Sophia or the divine feminine, Lucifer as the Illusionist, and Lucifer as Christ's brother in light—then we can see that here we have the best of relationship between Venus/Lucifer and the light of Christ expressing on the human plane.

---

45. Where the Sun stood at the Mystery of Golgotha, Venus stood at the event of Pentecost, when Mother Mary assumed the central role of mediator of the spirit of love to humankind. Sun at Mystery of Golgotha, 15 Aries 40; Venus at Pentecost, 17 Aries 2. These two events are joined in the Sun–Venus conjunction of this conversation.

46. See William Bento's article on "prophecy," op. cit.

47. Paul & Powell, *Cosmic Dances of the Planets*, op. cit., p. 145.

Summary: Prophecy listened to becomes intimate sharing in destiny. The destiny links male and female, heavenly destiny and earthly story. Venus listens.

1B: Related Event: Inferior Conjunction in the Twins: Death of the Solomon Jesus

On June 5 CE 12, the Solomon Jesus passed over the threshold, with Venus very close to the Sun.[48] This body was the vehicle for the great being Zarathustra, the most wise, who had transferred his wisdom to the one we think of as Jesus a few months previously, on April 3. On this day with Venus in inferior conjunction to the Sun, the bearer of wisdom from across the centuries died. Several themes interconnect in Bergman's film of *The Magic Flute* by Mozart (chapters 6 and 9), showing Sarastro (the same as Zarathustra, mentioned in the successful conclusion of his mission here, as well as in chapters 2, 3, 4, and Appendices B and D), reading Eschenbach's *Parzival* (taken up in #5 in this chapter, and all of chapter 8), before turning away the Queen of the Night (Venus in her dark aspect, the seductress that we find in Wagner's *Tannhäuser*, in #5D in this chapter).

Summary: Once the gift of wisdom has been conferred to one's successor within the karmic group, the bearer of wisdom can depart. One's duty to one's group, in this case all humanity, comes to an end, and one affiliates again with one's celestial group. Death occurs at the end of one's unique life purpose.

1C: Event: Superior Conjunction in the Lion: "I Have a Dream"

On August 28, 1963, Martin Luther King, Jr., delivered a speech that has echoed ever since as a poetical clarion call to social justice. He imagined

---

48. Venus at 15 degrees and 54 minutes of Gemini, very close to the Sun at 14 degrees and 12 minutes of Gemini. *Chronicle*, op. cit., p. 88, a reasoned estimate based on little evidence, and thus to be held lightly. Also, Mars was conjunct at 16 Gemini 16, and Neptune opposed at 16 Sagittarius 11.

in public how the different streams of brothers and sisters in the Venus sphere could come together: "I have a dream that one day on the red hills of Georgia, the sons of former slaves and the sons of former slave owners will be able to sit down together at the table of brotherhood."[49] One could see this in the head-pole or the will-pole, for King worked both ends. Whenever possible, he exhorted people to action, and this speech works with Venus at the will-pole.[50]

Summary: A remarkable man sets out a grand vision of the common destiny of humanity, where all join together as brothers and sisters. Venus listens and speaks; many listen. King orients others to their soul's yearning for community in service of humanity. This speech announces a karmic stream, to which some continue to feel called. King had the ability to mobilize life forces (Eros) and put a positive phantasm in front of them, more attuned to the soul's development than to the pictures of hatred which were coming at him.

1D: Event: Inferior Conjunction in the Waterman: Birth of The Mother

Mirra Alfassa, who became The Mother to Aurobindo's ashram and thriving community at Pondicherry, Auroville, was born in an inferior conjunction of Venus and the Sun.[51] The power of the feminine, the divine Mother, and thinking about the true destiny of the human body connect her

---

49. Venus at 10 Leo 1, Sun at 10 Leo 24.
50. King's birth Venus, at 16 Aquarius 49, lay square to the Aldebaran–Antares axis in Taurus–Scorpio. Another event that laid out an entirely new vision for humanity occurred in a superior Venus–Sun conjunction, the signing of the Magna Carta, giving rights outside of nobility, June 15, 1215, Venus at 16 Gemini 55 and Sun at 16 Gemini 24.
51. The Mother (Mirra Alfassa), was born February 21, 1878, both Sun and Venus at 9 degrees of Aquarius, an inferior conjunction, Venus between the Sun and Earth, though not eclipsing the Sun. Robert Lawlor, cited earlier in several places, was instrumental in changing the desert into the oasis of Auroville when The Mother ruled the ashram.

to the event of sharing between Christ Light and Mother Mary. Her major work had to do with a close listening to the human body, indeed an intensely intimate love of each cell of the human body. Venus listens and speaks.[52] She also held together a spiritual ashram which one could define as a karmic group finding each other and living together.

1E: Event: Superior Conjunction in Scorpio: Birth of Sai Baba

Sai Baba was born in 1926, very early gathering a karmic group connected with him personally, whose numbers grew to hundreds of thousands.[53]

1F: Event: Superior Conjunction in Scorpio: Birth of Mani

The great philosopher Mani was born with Sun conjunct Venus in the third century. As with The Mother and Sai Baba, he gathered around him a large group of followers connected by a feeling of joined destiny, a karmic group. We will speak about Mani in a different context in #5C below and in chapter 8.

2. Theme of Joining to Ancient Traditions

2A: Imagination: Inferior Conjunction in the Ram: Anointing for the Burial

Mary of Magdalene had enjoyed a special relationship to the teacher that the other students did not understand. Some of them resented her special position. When she brought one of her exotic perfumes to anoint their teacher, the men grumbled about the meaning of such behavior, Judas particularly. They wondered that the teacher tolerated this kind of thing. The day before, she had brought a bowl and washed his feet, drying them with a towel thrown over her shoulder, then putting a new ointment,

---

52. The Mother's thirteen volumes of writing are summarized in Satprem, *The Mind of the Cells or Willed Mutation of the Human Species* (Bend, OR: Institute for Evolutionary Research, 1982).

53. Sai Baba was born on November 23, 1926, both Sun and Venus at 6 Scorpio.

a new smell, on the teacher's head. On this day, she went further. She met him at the gate and unfurled her hair, wiping his dusty sandals with it. The men could not contain their grunts of disapproval, but said nothing as the teacher had not reacted negatively but had received her attentions quietly and then gone into the house. The men now settled in the room, and Mary Magdalene entered with a new ointment, rubbed it into the teacher's scalp and onto his feet, and then unfurled her hair again to wipe off the excess oil. The men spoke among themselves, agreeing that it was inappropriate for a woman to unfurl her hair, to use it to wipe away the excess—that the whole thing was entirely scandalous.

Comment: Most fresh in this story is the image of unfurling long hair and wiping up excess essential oils. The scents, the hair, the priestess—all of this lands us in the sensual realm of Venus.

This inferior conjunction of Venus occurred on March 21 CE 33.[54] Venus conjunct the Sun lay directly across from the Goddess star, Spica, the main star in the constellation of the Virgin, one of the Royal Stars of Persia.[55]

In the last weeks of the Teacher's ministry, Mary of Magdalene used seven ointments in all. The two ointments spoken of during this conjunction with Venus were second and third in the series. She used three vials at the Last Supper, where her behavior was questioned aloud—by Judas—and defended by the Teacher. Use of various ointments, especially the numbers seven or fourteen, formed part of the Egyptian system of preparation for death. Mary had been trained in that system. Here we see a snippet of a

---

54. Powell, *Chronicle,* op. cit.
55. The four main Royal Stars lie at the centers of the fixed signs, The Bull (Aldebaran), The Scorpion/Serpent/Eagle (Antares), The Lion (Regulus), and The Waterman (Fomalhaut). This fifth star, Spica, is semi-square (45 degrees) from the main axis of Aldebaran–Antares.

connection to the goddess traditions, fired by the relation to Venus. As we see the theme of Venus and one's karmic or spiritual group, we can see that Mary Magdalene declares Jesus as belonging, at least in part, to the great initiation traditions of Egypt.[56]

The sign of the Ram in this Sun–Venus conjunction evokes the virtuous quality of devotion as a power of sacrifice. Sacrifice is made on behalf of the group in service to which several groups are joined: Egyptian priest-craft through this priestess, the simple untrained folk who follow Jesus. This deep gesture of serving shows a chasm full of feeling. With Venus conjunct the Sun, directly across from the Goddess star, Spica, Mary Magdalene stands in for Sophia herself. Many have assumed that the con-nection between Magdalene and Jesus was sexual—an aspect of Venus—the unfurling of the hair, the familiarity with perfumes and potions, her capacity of sensual experience of the world. Sensual, yes, erotic, yes, but more accurately Magdalene's capacities demonstrated a refinement of Eros, expressing here as sensual devotion. She had punched through the phan-tasms and befriended Eros, aspect of Venus, the Goddess of Love, which stimulated her remarkable devotion. One cannot understand the pro-fundity of the meeting between Magdalene and the risen Christ without understanding this sense of devotion, active and fired in the head.

Summary: The inspired priestess links groups through intimate sharing of secrets. Eros becomes devotion.

---

56. The "day before" is March 19, and the day of the unfurling of the hair March 20 CE 33, Venus and Sun at 1 Aries 28, opposite Spica, the Goddess Star at the end of Virgo. The one-degree orb goes from late at night on March 20 to dawn on March 22. The understanding of Magdalene as a priestess in the tradition of Isis is presented by David Tresemer, in the preface to Jean-Yves Leloup, *The Gospel of Mary Magdalene* (Rochester, VT: Inner Traditions, 2002), and artistically in Laura-Lea Cannon and David Tresemer, *Rediscovering Mary Magdalene* (DVD from www.DavidAndLilaTresemer.com).

2B: *Event: Birth of Amedeo Modigliani, artist of the divine feminine*

Modigliani's interpretive paintings of women are astonishingly potent, deserving his birth mention next to this recognition of the feminine power of Mary Magdalene.[57] A lover of Venus can rejoice upon coming into the presence of one of his paintings. He knew the whole gamut of the expressions of Venus, and his paintings reveal a modern devotee of the divine feminine.

2C: *Imagination: Inferior Conjunction in the Ram: Cursing of the Fig Tree*

Walking with his students, the teacher stopped at a fig tree. It had leaves but no fruit. Jesus cursed it, saying the fig tree should not bear fruit any longer. The next day the students passed that way again, and were surprised to see that the tree was entirely barren of fruit and leaves, and that its branches had withered, becoming dry and brittle, some broken.

Comment: Many commentators have wondered about the significance of the cursing of the fig tree, especially so close to the Mystery of Golgotha.[58] It has been suggested that the fig represented the old religion, and that the teacher was announcing—and demonstrating—a new relationship with Divinity.[59] Speaking in terms of Venus' connection with karmic groups, the demonstration communicated a severance of connection with the old group and an announcement of the new. One does not affiliate with all ancient traditions but only with the ones to which one is connected by destiny.

---

57. Amedeo Modigliani, born July 12, 1884, Venus (26 Gemini 47) in inferior conjunction to the Sun (27 Gemini 11).
58. March 20 CE 33. Venus at 2 Aries 10 and the Sun at 0 Aries 23. Mark 11. See also Luke 13:6–9.
59. Among others, Adriana Koulias, in *The Secret Gospel* (forthcoming), develops the notion of the Fig School as the older religion, rejected by Jesus in this conjunction of Venus and Sun.

3. THEME OF RELATIONSHIP BEYOND HUMAN TO THE DIVINE

3A: Imagination: Inferior Conjunction in the Ram: Theophany

In the last days of his teaching, Jesus entered the temple, and said: "This day my soul is troubled. Shall I say, 'Father, save me from this hour'? No, it is my destiny to come to this hour. Father! Father! Glorify your name!" Immediately the sky thundered, in which many heard the words in the rumbles, "I have glorified it, and I will glorify it again!" When the thunder calmed, Jesus explained, "This voice has come for your sake, not for mine. When I am lifted up from the Earth, I will draw all people to myself." The people swooned with the impact of this demonstration of unity and the promise of the resolutions of death and time. From the room rose a unified exhalation, "Ah!"

Comment: Here Jesus affirms the importance of human destiny, and the affiliation with one's true group of belonging, represented by his own being. The theophany (meaning appearance of deity) affirms these truths through thunder and words.[60] In the Ram, we again find the sense of sacrifice of oneself to that which is greater. Sun and Venus lie directly opposite to Spica, the Goddess star, the star of Sophia. That is, the alignment here looks like this:

Beginning of the Ram (Aries)
—Sun—Venus—Earth—Spica—
end of Virgo

The vowel sound "ah," thought to arise from the planet Venus, could be uttered at every one of the theophanies, as well at most of the events mentioned in this book.

Given that this event occurred with Sun conjunct Venus, we can see it both ways, as a large grumble in the thunder from a fatherly divinity, and also as a singing at all frequencies from a feminine divinity, affirming the life forces moving in the world.

---

60. John 12:27–32, Venus 2 Aries 4, Sun 0 Aries 33.

Summary: Divine action engenders the ultimate group identification, the all with the All. Venus listens—to the voice of Divinity.

3B: Events: Superior Conjunctions: Divinity Speaks

At the two times that Divinity spoke previous to that which we have recorded, Venus lay in relation not to the Sun but to a powerful polarity in the cosmos, Aldebaran–Antares, 15 Taurus and 15 Scorpio, dividing the heavens into halves. On September 23 CE 29, Venus lay at 15 degrees and 3 minutes of Scorpio, at the event of the Water Initiation (Baptism), conjunct Antares. On March 28 CE 31, Venus lay at 14 degrees and 1 minute of Taurus, conjunct Aldebaran, the Life Star, when Divinity spoke in words that people could hear. Like the Sun, other major stars in the cosmos can magnify the influence of a planet, and these two greatest of the Royal Stars of Persia are known as the hinges of heaven, around whose polarity the zodiac was organized. Thus the event of the speaking of Divinity is shown to relate to strong positions of Venus, where she is empowered by the Sun or great stars. Venus listens.

3C: Event: Inferior Conjunction in the Crab: Historical Event:
The Entry of the First Crusaders into Jerusalem

On July 15, 1099, heeding a call to reclaim their connection with the life and vitality of Jesus Christ, the first Crusaders accomplished the goal set many months before: They entered Jerusalem to meet "the bride beautifully dressed."[61] This rejoined the masculine warriors with the feminine home, as bride and as mother. For all of European Christendom, it reconnected the believers with their karmic group and karmic geography, karma here meaning the influence that destiny has on one's life beyond this day and even this lifetime. Was it karmic necessity to have Western ownership in this multicultural center? The Crusaders were, like Judas (see below), propelled

---

61. Reve. 21:2; also 21:9.

by a vision (head-pole) of what they wanted, perhaps an early version of the Jerusalem Syndrome, where a kind of messianic hysteria comes over otherwise sane and stable people.[62] Rather than taking the experience into their hearts, and negotiating truces along the way, they wanted exclusive possession, and murdered to achieve their goal.[63]

Summary: Christian culture yearned to lay claim to the geography of the Christ events. They yearned to reunite with the karmic group of the origins of their religion. Zeal to connect with one's destiny group leads to extreme deeds.

3D: Event: Superior Conjunction in the Bull: Johannes Kepler's Discovery of the Third Law of Planetary Dynamics

At a superior conjunction of Venus and the Sun (connected with the will-pole of the human being), Johannes Kepler discovered the way in which the heavens were ordered according to musical harmonies.[64] Here a daring human being looked out beyond the thought forms of the day, peering into the mysterious dynamics of the heavens, and found harmonic order. The theophany appeared to Kepler as the square of the period of rotation of a planet around the Sun in proportion to the cube of its distance from the Sun.

---

62. Though disputed as a technical diagnosis, you get loads of examples whenever you ask anyone from Israel. The defining study is Y. Bar-el et al., "Jerusalem syndrome," *British Journal of Psychiatry*, 2000, pp. 176, 86–90.

63. July 15, 1099, Venus 16 Cancer 27 in inferior conjunction to the Sun 15 Cancer 33. At another dramatic battle where displaced people came in force to reclaim a connection to their land, namely on D-Day, June 6, 1944, Venus (at 15 Taurus 55) lay conjunct Aldebaran, the Life Star, an influence very powerful in the heavens (chapter 5). This all has a shadow side, too. The beginning of World War II, just as its end, was related to a Venus event: Germany invaded Poland because it was claimed to belong to the Fatherland, September 1, 1939, Venus 12 Leo 38, Sun 13 Leo 55. About another beginning to the war, Pearl Harbor, as well as the end on D-Day, see chapter 5.

64. May 15, 1618 (Venus 6 Taurus 35, Sun 4 Taurus 59). We have this date from Nick Kollerstrom's historical research in *The Eureka Effect*, both book and website.

Kepler wrote in Book V of *Harmonices Mundi*, in introduction to this third law: "I have stolen the golden vessels of the Egyptians from which to furnish for my God a holy shrine far from Egypt's confines." He thus connected the spirit/karma stream of the Egyptian initiates (see again Imagination 2 above) with the present. The golden vessels can remind one of the golden apples of Venus mentioned earlier in this chapter, and of Jamshid's great cauldron.

Kepler continued, "But now, Urania, there is need for a louder sound while I climb along the harmonic scale of the celestial movements to higher things where the true archetype of the fabric of the world is kept hidden."[65] One could say that Kepler felt he was lifting the veil of Isis/Sophia/Venus/Divine-Feminine by entering through music—and the music of the planetary orbits.

Kepler's accomplishment was less a feat of the mind than an ascent of a celestial ladder, a theft of golden vessels, an act of will to share heavenly secrets with brothers and sisters.

Summary: The Divine expresses as harmonious musical order, and a human can perceive this beauty.

3E: Events: Other Theophanies

The first man thrown into space might be a modern scientific/materialist world version of this connection of the community of humanity with the All: Yuri Gagarin pressed through the atmosphere into the heavens on April 12, 1961.[66] It is interesting to conceive of this as more related to the head-pole, to how we think about ourselves, than as an act of will.

The first explosion of an atom bomb occurred on July 16, 1945, Venus conjunct Aldebaran, whereupon the head of the project to develop this

---

65. Kepler invokes Urania, the muse of astronomy. Though the planet Uranus had not yet been identified in the heavens, on the day of Kepler's discovery of the third law, December 27, 1571, Uranus (26 Gemini 37) lay directly opposite Kepler's Birth-Sun (26 Sagittarius 41) and Venus (29 Sagittarius 48), thus complementing them.
66. Venus lay at 26 Pisces 0 in inferior conjunction to the Sun at 27 Pisces 42.

penetration of the atom's secrets, the physicist Robert Oppenheimer, felt the revelation of Krishna as ten thousand suns, recorded in the *Mahabharata*. Peering into the immensity of light, Oppenheimer heard and spoke the words, "I am become Death, destroyer of worlds."[67] Something fundamental was revealed about the foundations of our existence as expressed in pure energy, Isis unveiled, an experience of her naked power. From this we can see why Venus (in her guise as Isis) must normally be veiled—the ancient warning goes "And never shall mortal man perceive me as I am"—and why so many myths picture the one who beholds her unveiled burning up.

The birth of Leo Szilard, the man responsible for the chain reaction experiment described in chapter 5 for the 20–21 degree of Scorpio, was born with Venus conjunct the Sun, another connection between Venus (as Isis) and the destiny of the world.[68]

Finally, the invention of the particle accelerator by Ernest Walton—permitting a look into the insides of the atomic world—occurred when Venus lay conjunct the Life Star, Aldebaran.[69]

4. Theme of the Personal Call to One's Karmic Group

4A: *Imagination:*
The Call to Join the Brotherhood, Inferior Conjunction in the Goat

The teacher walked purposefully, and his students walked swiftly to keep up. Each wished to stay close, to hear every word that the teacher might utter. They had traveled over many kinds of terrain and through many kinds of weather, and the students felt intimate with the teacher. One said as they

---

67. Venus at 15 Taurus 14, conjunct Aldebaran, the Life Star. Conjunctions with the fixed stars are always inferior conjunctions.
68. Leo Szilard was born February 11, 1898, both Sun and Venus at 29 degrees of Capricorn, superior conjunction.
69. This occurred on April 13, 1932, Venus at 15 Taurus 7, conjunct Aldebaran.

passed by the boats on the shore of the great lake, "Teacher, look there, many fishermen caring for their boats. We know them! They should come join us, and spread the words of your teaching. Master, shall I go beckon them to join us?"

The teacher paused and observed the scene—the boats, piles of nets, huts for equipment necessary for repairs, and all the activity of men, women, and children to support the task of catching the fish of the great lake. "I have not yet called them, though I will do so. What they do supports many people, and they must do that service for now, until they are needed."

Comment: This occurred at an inferior conjunction of Venus to the Sun on January 8 CE 30, with Sun and Venus at 20 degrees and 36 minutes of Capricorn.[70] It demonstrates Venus' connection with membership in groups, who will be in and who will not be in. The fishermen observed included Peter and Andrew. Our connection with a group wherein we are seen, wherein we develop relationships of love to our brothers and sisters, and to which we can freely give all of our heart and strength, is very important to our soul's development.

Summary: Venus orients us to the groups to which we belong, as well as to our brothers and sisters, and to the right time to admit them.

4B: Imagination:
The Call to Judas and Thomas, Superior Conjunction in the Scorpion

Bartholomew pointed out a man in the distance and told the teacher, "Look, there is a man who is earnest in his seeking, a hard worker, modest in his appearance, trained in his uncle's business of tanning hides, and good with numbers. His name is Judas, from Iscariot. Though his mother has a bad reputation, and his father is not known...."

---

70. All of the dates for these events come from Powell's *Chronicle,* op. cit.

Another student spoke up sharply, "His mother is a dancer of exotic dances, who has had many men." He waited for the expected mumbles of concern among the other students. "She abandoned her child when he was an infant, near a waterhole where wild animals were known to come." More mumbles. "She has no idea of who the father was." This was too much, not knowing your parentage and therefore your blood group. Several students exclaimed their amazement that such a person was even being considered.

Bartholemew defended Judas, "I know him. He has talents and good qualities. His uncle raised him and can vouch for him. Teacher, looking at the man himself, do you not think that he would make a good student and be helpful to us?"

The teacher sighed and said, "Yes, but not yet, not just yet."

Two days later, the teacher and Judas met. Judas asked to become a student, to study the teachings diligently. The teacher said, "There is a position as one of my students. You can take it. Or you can decide to leave this position for another." Judas insisted that he would join, for he wanted to be part of the most exciting movement of his time.

A week later, the teacher met with Thomas, another young man who had been recommended, a student of law, who had been trained in doubting anything based only on hearsay unless evidence was presented. Thomas asked to join the brotherhood and, more than that, the community of brothers and sisters that had gathered around the teacher. The teacher agreed.

Comment: These calls occurred during a superior conjunction. On October 24 CE 30, with Venus at 2 degrees and 19 minutes of Scorpio and Sun at 1 degree and 50 minutes of Scorpio, Judas was given the choice to come on the path that the teacher saw lay ahead.[71] On October 29, the

---

71. At this meeting the Sun lay on the great circle (perpendicular to the ecliptic) that included Algol. This meeting is discussed in those terms in William Bento, Robert Schiappacasse, and David Tresemer, *Signs in the Heavens: A Message for Our Time* (from www.StarWisdom.org, 2000).

teacher summoned Thomas, the student of law, the doubter. The other disciples (that we know about) were chosen in connection with other planets.[72]

The choice of both Judas and Thomas brought into the group two of the most willful of the lot, a sense of the superior conjunction relating to the will-pole of the human being. These men were responsible for deeds. Judas' deed of betrayal (#4C) was absolutely necessary to the Teacher's unfolding story. Thomas acted out the demand for physical proof, needing to put his fingers (extension of the will-pole of the arms) into the wounds of the risen Christ.

The mention of Judas' mother as an exotic dancer brings home the basest expression of Venus, alerting us to the will-pole of Venus conjunct the Sun.

Summary: One hears the call to one's karma or destiny group. Others discuss you—the whole group is involved.

4C: Imagination:
Judas Shifts Allegiances, Inferior Conjunction in the Ram

Judas had come back with the students and the teacher through the Golden Gate in a triumphant procession. But he was dissatisfied, and when the rest were busy with meals and plans to distribute goods to the poor, he slipped out and ran to the house of Caiaphas, the high priest of Jerusalem, and agreed to identify which of their band was the leader, Jesus.

Comment: Judas takes "the first definite step in his treacherous course," that of exposing the teacher in order to force his hand to demonstrate his

---

72. Thomas was chosen with Venus at 8 degrees and 31 minutes of Scorpio and Sun at 6 degrees and 49 minutes of Scorpio. This is a bit greater than our usual orb of one degree, but obviously relevant to the issue of being called to the group to which one belongs. Andrew was chosen at a moment when the Sun lay opposed to Saturn and conjunct to Pluto; Matthew was chosen with Sun square Mars; Peter was chosen with Sun square Spica; Philip was chosen with Sun conjunct Mercury and square Spica. The timing gives each of these students a particular celestial imprimatur that colors their participation in the teacher's community. Recall that Andrew and Peter had been among the fishermen when they were considered in the Venus–Sun conjunction of #4A, but not found ready quite yet.

powers by overthrowing the Roman occupying army.[73] In this act Judas dis-affiliates himself with one group and affiliates himself with another, con-tinuing the theme of calling and belonging. He is propelled by his vision of what Jesus should be, thus a relationship to the head-pole of the inferior conjunction. Some say that his new group of affiliation was the Romans, some say the messianic overthrow of the Romans, and some say his new group included all those who needed Christ Jesus to go through the Mystery of Golgotha for the sake of world evolution.

Summary: One tries to force one's destiny group to change, based on one's own ideas and illusions.

4D: Imagination:
The Call to Zacchaeus, Superior Conjunction in the Twins

The teacher led a band of several hundred as they entered the city of Jericho. Crowds pressed in from both sides of the street, cheering, the healthy and the sick. Nearly all the way to the town square, the teacher stopped and turned to face one of the great sycamore trees that lined the road into the town. He held up his hands, and the crowd grew silent. He shouted up to the tree, "Zacchaeus!" There was complete silence. "Zacchaeus!" Some leaves shook, and little Zacchaeus, the greatly feared tax collector and an agent of the Romans, stuck out his head. The crowd gasped. The teacher held up his hands again. "Zacchaeus, come down. Prepare your house, for today I shall enter." The teacher then turned and waved for all to follow him to the town square. The entire popu-lace cheered and fell in behind him, knowing that there they would find the healing that they so earnestly sought. Later Zacchaeus changed his allegiance to the tax collectors and the occupying forces of Rome to the brotherhood and sisterhood of Jesus.

---

73. March 20 CE 33, inferior conjunction.

Comment: This encounter is developed in a story for Mars, as there is a significant aspect with Mars here, too.[74] This superior conjunction of Sun and Venus on May 30 CE 32, with Venus and the Sun at 8 degrees and 8 minutes of Gemini, joins the call to Zacchaeus with the other calls to join the community of destiny.

Summary: In the sphere of Venus, one hears a call to join a new group, entering affiliations and relationships related to one's destiny.

5. THEME: THE QUEST OF PARZIVAL

We devote chapter 8 to the connections of Parzival to the themes of the Venus eclipse of 2012. Here are preludes in the context of Venus.

5A: Event: Superior Conjunction in the Crab: The Tale of Parzival #1, First Visit

The tale of Parzival tells of the Grail King's wound that will not heal, and the young man awaited to solve the enigma and heal the old king. Parzival, the young innocent, chances upon the Grail Castle and meets the Grail King, and observes a ceremony of profound meaning, to which he feels drawn. Confused, young, trained to hold his tongue, he says nothing. He listens. He does not ask about what is going on. Specifically, he does not ask the question for which the King and all the court have waited. He leaves the next day, and learns of his inadequate behavior from Cundrie. She is the counterpart to Mary Magdalene in #2A, both acting as the divine feminine, Venus, in relation to world events. Parzival must search for years before he can find the Grail Castle again and ask the special question that heals the wounded King.[75]

---

74. I feature this story in my presentation of the signature of Mars (forthcoming).
75. The first visit of Parzival can be seen to occur on July 15, 828 (Julian), with Venus at 15 Cancer 57 and the Sun at 17 Cancer 57. The dating has been worked out by Joachim Schmidt, in an article written in 1947, "Parzival and the Stellar Script," published by Suso Vetter in the 1987/88 edition of the *Sternkalender* (star calendar), published by the Philosophisch-Anthroposophisch Verlag, Dornach).

Summary: In the aura of Venus, one can observe one's destined group as a mysterious stranger, then be lost from it. One listens but does not speak—too much listening. "In this learning to ask questions lies the ascending stream of humanity's evolution."[76]

5B: Event: Superior Conjunction in the Bull:
The Tale of Parzival #2, Completion

After five years of wandering and learning through life experience, Parzival is ready to return. He earns the role of Grail King near Easter, a deed that is sealed at Whitsun or Pentecost fifty days later, another Venus–Sun conjunction (this time at the degree of the Venus eclipse of 2012).[77] In the Bull this completion at Whitsun refers to the deeper cosmological significance of that constellation in relation to Earth evolution—the foundation of the "I"! It opens up the new mystery schooling from an esoteric Christian perspective—the new gathering of souls in disciplined devotion to the center, the Grail, in an important way the new yoga.[78]

Summary: In Venus, one finds completion of one's destiny in leadership of a group of others to find their destiny.

William Bento works with this event and date in "Saturn in the Crab and the Mysteries of the Holy Grail," *Christian Star Calendar, 2006* (the immediate predecessor of *Journal for Star Wisdom*). At the event in the year 828, other planets were conjunct to Venus and the Sun: the Moon at 14 Cancer 24, Mercury at 21 Cancer 45, Jupiter at 13 Cancer 52, and Saturn at 22 Cancer 51. The Sun in 828 lay at the place where Saturn lay at the Mystery of Golgotha (18 Cancer 45). More about this can be found in the wonderful publication by Robert J. Kelder (ed.) *Werner Greub: Wolfram's Grail Astronomy* (Willehalm Institute for Grail Research, 1999, also available from www.StarWisdom.org).

76. Rudolf Steiner, *The Fifth Gospel,* cited in Selg, op. cit., 113–114.
77. May 15, 833, Venus at 19 Taurus 51, Sun at 19 Taurus 30, using Joachim Schmidt's dating.
78. See Walter Johannes Stein, *The Ninth Century: World History in Light of the Holy Grail* (London: Temple Lodge, 1991).

5C: Event: Superior Conjunction in the Ram: The Tale of Parzival #3: Birth of Mani

The philosopher Mani, founder of Manichaeism, was born with Venus in superior conjunction to the Sun.[79] He has been considered a previous incarnation of Parzival.[80] His philosophy linked together many of the karmic streams governed by the sphere of Venus. Manichaeism can be found in the Templar Knights (cf. the entry to Jerusalem above), the life and accomplishments of Parzival, also in the Cathars and other mystical spiritual streams that emphasized the development of the individual's relationship to Divinity, and the sense of uniting in the light to confront strongly the dark.

Summary: In the sphere of Venus, brothers and sisters that share a human initiative find each other. Different expressions of the same individuality link up across time.

5D: Event: Superior Conjunction in the Bull: The Tale of Parzival #4: Birth of Richard Wagner

The great composer Richard Wagner was born with Venus conjunct the Sun.[81] His music was felt as the most powerful expression of the German and indeed European folk soul, and united millions of people in the sense of a shared karmic group. His last opera, *Parsifal*, first performed in 1882, links him with the theme of Parzival. It pictured groups

---

79. Mani, born April 14, 216 (J), Venus 24 Aries 0 and Sun 23 Aries 12. Richard Seddon, *Mani, His Life and Work, Transforming Evil* (London: Temple Lodge, 1996); L. J. R. Ort, *Mani: A Religio-Historical Description of His Personality* (Leiden, Netherlands: Brill, 1967), p. 156.

80. Seddon, ibid., chap. 9. The reincarnation of Mani to Parsifal is suggested by Rudolf Steiner. See Sergei O. Prokofieff, *Relating to Rudolf Steiner: And the Mystery of the Laying of the Foundation Stone* (London: Temple Lodge, 2008), p. 18.

81. May 22, 1813, Venus at 7 Taurus 33, Sun at 8 Taurus 26, also conjunct the ascendant at 5 Taurus 32, thus making the Venus–Sun conjunction especially important in the expression of his life. Uranus lay exactly opposite at 3 Scorpio 36.

of men and women, some lost, some ill, some enchanted, all finding each other again through alignment to the higher purposes in their lives, namely through the affirmation of the Holy Grail.

As prelude, however, to Parsifal, Wagner wrote the Ring Cycle of operas, which uncovered the dynamics of the heavens, gods and human heroes in conversations and conflicts. The themes of Venus, from the most sublime to the most base, weave throughout.

Most important for this study is another opera, *Tannhäuser*, first performed in Dresden in 1845.[82] It begins with the great singer Tannhäuser captive in the castle of Venus. He has been her lover for a time out of time. He sings his joy:

> Gratitude for your favor and praise for your love! Forever blessed is he who has dwelt here! Forever envied is he who, hot with desire, has shared, in your arms, the divine glow! The wonders of your realm cast a spell! Here I breathe the magic of unalloyed bliss. No land in the wide world offers the like. All it holds seems in comparison of little worth.

Then something wells up within him, the desire to escape the chains of Venus:

> Yet from these rosy scents, I long for the woodland breezes, for the clear blue of our sky, for the fresh green of our meadows, for the sweet song of our birds, for the dear sound of our bells. From your kingdom I must flee. O queen, goddess, set me free!

The conundrum is this: He seeks to flee from the bed of Venus into the world of Venus, the woods, sky, meadows, birds, sounds of bells.

Venus allures him: "Revel in union with love's own goddess." He succumbs, then rises up again: "For freedom I thirst."

---

82. *Tannhäuser and the Singers' Contest at Wartburg*, first performed October 19, 1845, with Venus at 13 degrees and 19 minutes of Scorpio, thus conjunct to Antares.

Venus mocks him: "Fly to the cold world of men, and their feeble, cheerless fancies." There are strong and cheerful fancies—phantasms—here in Venus' embrace. And phantasms await Tannhäuser there in the outer world, fancies that are cold, feeble, and cheerless.

The rest of the opera traces Tannhäuser's attempts to escape the deepest addictions to Venus' temptations. In the character of Elisabeth and Tannhäuser's relationship to Elisabeth, the opera moves through realms of love, devotion, forgiveness, the entire gamut of the continuum of Venus given earlier in this chapter. He is sent to receive forgiveness from the Pope for his sex addiction. In Rome he's laughed at, and told that he will be forgiven only when he has truly learned a higher love, at which point the Pope's staff will sprout new leaves, which no one believes will happen. He returns in rags, dies, and then the announcement comes that the Pope's staff has sprouted new leaves. As the proof of innocence lies still within Venus' command of the power of generativity, we have witnessed an entire opera about moving from one sort of relationship with Venus to another.

In *Tannhäuser* we see Wagner's familiarity with the encounter with Venus, and the investigation of the different sorts of relationship one can have with her.

5E: Event: Superior Conjunction in the Bull:
The Tale of Parzival #5: Birth of Trevor Ravenscroft

An author who popularized the tale of Parzival more than any other, in his books *The Spear of Destiny* and *The Mark of The Beast*, was born with Venus conjunct the Sun.[83]

---

83. Trevor Ravenscroft was born April 23, 1921, superior conjunction of Sun and Venus at 8 Aries. His books are cited in chapter 8.

In this chapter we investigated the character of the sphere of Venus by studying those people and events that express its vivifying influence. This involved Venus in conjunction to the Sun, either inferior (nearest to the eclipse phenomenon) or superior, as well as, in a few instances, in relation to Aldebaran or Antares (cited in the commentaries for 1A, 3B, and 3E). Pondering these signatures of Venus leads to a better understanding of Venus' role in the Venus eclipse of June 2012.

VENUS ECLIPSING THE SUN—BEAUTY AND TECHNOLOGY

How does the specific character of Venus relate to the alignment that we're pondering:

20–21 degrees of the Bull (Taurus)
—Sun—Venus—Earth—
20–21 degrees of Scorpio

Henry David Thoreau integrated some of the themes that we've discussed in *On Walden Pond*, which I've set out as a poem:

> To be awake is to be alive.
> I have never yet met a man who was quite awake....
> We must learn to reawaken and keep ourselves awake,
> Not by mechanical aids,
> But by an infinite expectation of the dawn.[84]

The first two sentences relate to the daily cycle of day and night, and the suggestion that everyone that he's met sleeps to some extent. We might say, they live within the thrall of erotic phantasms of one sort or another. The

---

84. This quote is printed and framed in the Concord Museum in Concord, Massachusetts, not far from Walden Pond, which has now become a park. Thoreau began his experiment at Walden Pond on July 4, 1845, his way of saying that his form of independent awakeness on Independence Day was individual sovereignty in the forest, with a bean field.

third sentence implies that we once were awake, and now we must learn how to reawaken, "by an infinite expectation of the dawn," meaning in relation to the rising of the Sun, part of the time heralded by the Morning Star, that is, Venus. The fourth line gives us a warning. In Thoreau's time, "mechanical aids" may have meant an oil-filled lantern, yet this quote relates much more to the technologies of our time and to the theme from 20–21 Taurus–Scorpio accentuated by the Venus eclipse. For Thoreau, not using an oil lamp meant that he did not miss the dawn of the day, the advent of the Sun, because it brought the yearned-for capacity to see what he was doing, to participate in the being and beauty of Sophia's creation.

Thoreau distrusted technology, noting that "men have become tools of their tools" rather than masters of their tools. His observations are extremely pertinent to the themes of the Venus eclipse of 2012.[85]

On a few occasions, the philosopher Rudolf Steiner spoke directly to the effects of technology that bring us back to Venus. In comments given informally at the end of one of his lectures, Steiner spoke of his relationship to motorcars, citing their utilities and their dangers. On the one hand, they are noisy, and they wrench us away from a relationship with the Earth. On the other hand, he admitted that he found automobiles useful to get from place to place. The key for these conveniences was balance: "One must not wish to arrest the world's development," meaning that we needn't push back technological advances, "but must balance what comes from the one side by something else coming from the other."[86] One imagines therapeutic

---

85. A ubiquitous quote, found widely, attributed to *On Walden Pond*. Thoreau had a most compelling connection to the 20–21 degree of Taurus/Scorpio, and thus to the event of June 2012. At his birth, the planet Uranus lay at 20 degrees and 11 minutes of Scorpio; at death the planet Uranus lay directly opposite at 21 degrees and 49 minutes of Taurus. His life of 45 years lay within a half-cycle of Uranus at the polarity amplified by the Venus eclipse of 2012.

86. This comment by Steiner on motorcars occurred at the end of the lecture at Penmaenmawr, Wales, on August 11, 1923, in Rudolf Steiner, *The Evolution of*

balancing of riding in cars with walking, much more walking that we're used to.

In this lecture he went a bit further. "With the gramophone it is a different thing." From gramophone to record player (large discs of vinyl and a needle resting on them) to cassette tape to CD to iPod, it is all the same thing. "With the gramophone man is trying to bring the mechanical into the realm of Art...so that what descends into the world as a reflection of the Spiritual is made mechanical." In that case, humankind "will be able to defend itself no longer, there will be no help to be found. Help can then only come from the gods."

Let's figure this out. When you drive a car, you have uppermost in your mind the transport of yourself from one place to another. You are interested in the utility of it; the service occurs in the area of utility for the will. When you enter into the realm of feeling, of the heart—of Venus in her finest expression—you become especially vulnerable to a connection with spiritual realms. The subtlety that comes through a Modigliani painting seen up close, for example, or a Mozart symphony heard live, nourishes your soul. You have the phantasm of the expression of beauty, and you can sense the Eros through the pigments of a painting or the tones of a symphony, through the painter and musicians, back to the sources of Eros itself. As such, these encounters with art sustain and deepen your life. Observe yourself when you listen to a concert via any of the many methods by which we play these things today. The phantasm is similar; indeed, one could say that it's better, because most recordings use several renditions of the same symphony and weave together the best parts to make the final recording. The phantasm is there, but the Eros is more difficult to find. You may notice that in a live performance, something flows into you, the warmth of the artistic

---

the World and of Humanity (Blauvelt, NY: Garber, 1989), pp. 200–201. Other comments were made on Aug. 27, 1923, and March 29, 1920.

presentation. And you may notice that in a canned performance, something flows in—stimuli, color sensations, sound sensations—but that something flows out of you too, your search for the Eros behind it. This soul-longing drains out of you. Usually you don't notice it, as you don't notice many of the movements of your soul unless you watch very carefully. You think you're satisfied with what came in through the phantasm. You've opened what Lucretius called "love's unhealable wound," expecting something to come through those deeper levels, which it does when experienced in the flesh through human musicians and with the radiant original paintings before you. But it hasn't come through and you've missed the Eros. That's why Steiner said that we are not defended, as this all happens typically below the surface of consciousness.[87]

I have noticed that people fill in the missing Eros for themselves. This is fine to do now and again, but we are meant to receive heavenly sustenance, which the technology prevents. I continue to listen to recorded music, and eagerly anticipate the opportunities to experience true art as live, in its beauty and enlivening of my heart and middle realm.

Steiner concluded his comment on recorded music on a hopeful note: "But the gods are merciful, and we today may hope that as regards the progress of humanity, the merciful gods will themselves help it [humanity] to overcome such aberrations as are expressed in the gramophone." He further suggested, as an antidote to these forms of technology, "we must possess hearts yearning for the spiritual world," that is, consciously and with intention.[88] Mechanical conveniences, if thought of as utilitarian, can be balanced by other activities. Those activities are a kind of defense against the erosion

---

87. Lucretius, *De Rerum Natura*, op. cit., Book I. Other media for the expression of beauty are different, of course. What is the best way to experience the true transmission of a poet? By hearing the poet herself or himself speak the poem? Yes. Or by reading it aloud yourself, with feeling.
88. Ibid.

of our connection to Sophia. However, "mechanical aids"—Thoreau's term again—that present themselves as art, in the realm of beauty—can wrench us away from our connection to the life springs of creation, can wrench us away from our life's intention, can wrench us away from the soul groups to which we are linked by karma, and to which Venus has directed us. And there's nothing we can do about it after we have given away our sensibilities for beauty to the mechanical.

From the point of view of the realm of Venus, the emphasis of this chapter, we see that technology, though it may come from the working of the will, affects the middle realm—the realm of Art, beauty, grace, and the heart—in a special way. Steiner's warning that we may become defenseless and lost reemphasizes the importance of inquiring into the effects of technology, of noticing when it is a servant of convenience, and when it tries to take on the middle realm of the human being, the realm of art and of the heart.[89]

---

89. I have not included other perspectives on the conflict between true Eros/Venus and technology, but could mention two others here that could deepen the interested reader's understanding. Herbert Marcuse offers a brilliant analysis in *Eros and Civilization* (1955) that balances Freud's pessimism (more akin to Bruno's) in *Civilization and Its Discontents* (1929). Second, one could tie the case for civilization as ordered by Jamshid's tools with the Apollonian approach, in contrast to the more lively, erotic Dionysian approach, a polarity explored by Friedrich Nietzsche, Ruth Benedict, and others. Marcuse, along with Wilhelm Reich, defended the beleaguered erotic against the repressions of the technocrats. These thinkers assist only to the extent that you can find ways in which they apply to your personal life. Becoming too philosophical about these (re)pressing matters distances one further from the root sources of Venus' power.

8

PARZIVAL AND DEATH

We have seen Parzival come up several times so far in this study, especially chapter 7, #5. Let us now examine the story for its connections with the Venus eclipse of 2012, as a springboard for further inquiry into why this tale is pivotal.

THE STORY OF PARZIVAL

The story was told by the bard and troubadour Wolfram von Eschenbach in the early thirteenth century.[1] It begins with Parzival's parents, giving many details about places, the genealogy of all the actors and their escapades, for example, the death of Parzival's father in a distant land while adventuring as a knight. Parzival's mother rejected the culture of war and took the young boy into the forest, where they had no weapons. Parzival lived a technology-free life. His mother shielded him from the experience of death.

One day as Parzival was wandering among the trees, three knights on horseback came by, asking the boy if he had seen the one they were pursuing. The sunlight played on their armor and on the richness of their technological

---

1. The unfinished version by Chrétien de Troyes was written in 1180, titled *Perceval: Le Conte du Graal*. Wolfram von Eschenbach wrote in the early thirteenth century: *Parzival: A Romance of the Middle Ages*. The anthroposophic movement and Waldorf schools favor the translation by Helen Mustard and Charles Passage (New York: Vintage, 1961). A new and brief version was written and lavishly illustrated with dramatic pastels by David Newbatt as *Parzival* (Stourbridge, UK: Ruskin Mill, 2003). Based on this tale, Linda Sussman wrote *Speech of the Grail: A Journey Toward Speaking That Heals and Transforms* (Hudson, NY: Lindisfarne Books, 1995) .

gear. Parzival fell to his knees and asked in a trembling voice, "Are you gods?" Thus would Jamshid be treated by anyone who saw him with his magic tools. We wonder if our magical tools make us gods in the eyes of others.

Parzival returned home insistent upon becoming just like the knights he had met in the forest. His mother, saddened by the breakdown of her protections, tried to dissuade him, but he persisted. She sent him off with instructions to listen and not to speak, and to treat all ladies with kindness, and other rules of chivalric behavior. Venus listens, and she also speaks. At this early stage, Parzival was trained to honor relationship by listening, but not to reciprocate. This inequity would prove disastrous later on.

He then had a series of adventures and misadventures, making mistakes, some humorous and some grievous errors, learning along the way how to be a knight. He entered the world of knights who fight each other at small provocations, jousting with long lances which splinter—it is repeated that in an encounter the combatants used a forest of lances, splintering them all—and then with swords, until one was killed or, more often, surrendered. The defeated one was then sent, on his honor, on an errand by the victor, usually to take a message to a lovely lady in some distant place.

At one point, Parzival entered the Grail Castle, in which he observed the agony of the King of the place, who had a wound from which he could not heal. He observed the ceremony of the Grail itself, brought forward by twenty-four maidens. This recalls Jamshid's great cauldron, for the Grail has magical powers of healing. When the Grail touched the wound of the King, he was brought back from the edge of death, though continued to suffer from his wound. The wound had been inflicted by the magician Klingsor with the lance used at Golgotha to pierce the body of Christ Jesus. The wound was to the genitals. The spear was a version of Jamshid's golden dagger.

Parzival observed the wonders of the Grail Castle, not realizing that he was related to everyone there, and not realizing that he was supposed to do more than listen. He was supposed to have asked The Question, "What ails thee, uncle?" In old German the expected question begins, "Oeheim," which is a familiar term for a relation such as an uncle. It continues, "was wirret dir?" This can mean "Dear uncle, how are you confused? What is in you that makes you susceptible to error? How are you going astray? Dear uncle, what's it like inside of you that you can tell me about?"[2] Such questions encourage sharing, and are thus part of the stance of Venus' listening.

That question would have healed the Grail King. This first encounter took place with Venus in superior conjunction to the Sun (from #5A, chapter 7).

Parzival awoke the next morning to find the castle empty. When he saddled his horse and rode out, he received a stinging rebuke from the feminine figure Cundrie, who told him that he had failed. What he had truly failed at was recognizing his karmic brothers and sisters. This message was delivered by a Venus figure, sometimes beautiful and sometimes ugly.

Parzival wandered. When in the forest, three times he met a woman mourning over a dead knight. The knight slowly decomposed, and finally his partner died too, both in increasing purity.[3]

---

2. Often translated simply as "Tell me, Uncle, what troubles thee?" or "Uncle, what ails thee?" See Walter Johannes Stein, *The Ninth Century and the Holy Grail* (London: Temple Lodge, 1991). Modern speaking emphasizes the "what," as in asking for a diagnosis, rather than a more intimate expression of warm interest about the interiority of the King. A modern might say, "So what's your problem?!"—thus looking for the "thing" that is outside the victim which has been acting as a thorn that can be removed. Hans Mahle has translated this phrase: "Uncle, how have you become so entangled?," emphasizing the true interiority of the experience. We add *dear* in "dear Uncle" because *dir* is an intimate form of "you."

3. The knight was Schionatulander, who has been identified as a previous incarnation of Rudolf Steiner. Schionatulander had died protecting Parzival, though Parzival didn't learn this until much later. Thus Steiner had a strong hand in this development of Venus and the Sun through Parzival's development, in particular concerning the issues of the Venus eclipse of 2012.

Parzival wandered further. Meanwhile the story followed his alter ego, Gawain, who took on the magical technology of Klingsor's Castle of Wonders. The black magician Klingsor had powers of technology to ensnare knights and ladies into his castle to serve him. He sent out young women, made beautiful by his magic, to seduce strong men, manipulating Venus' power. Technology and sexuality intermixed. Once enchanted, both men and women moved as in a daze, serving this dark lord.

Years passed. Parzival became discouraged in his quest for the Grail Castle. He met the hermit priest Trevrizent, who taught him about the dynamics behind these family stories, and about the life and drama of Christ Jesus that lay behind it all. For it happened that the day they met was Easter, and, based on Trevrizent's guidance and the maturity he had gained from his many experiences, Parzival was able to find the Grail Castle that very day. At the end of the drama, Parzival asked the question of the wounded Amfortas. "Oeheim, was wirret dir?" That kind of warm interest activates a soul and begins a true conversation. Parzival asked not about the "thing" making Amfortas ill but about his interior activity and structure, about his Venus nature.

Parzival was crowned Grail King, a coronation sealed by the subsequent Pentecost, when Venus lay conjunct the Sun at 19–20 degrees of the Bull, just minutes away from the location of Sun and Venus at the eclipse of 2012 (#5B, chapter 7). The completion of the long adventure occurred in consonance with the Images of the Bull and of the Scorpion explored in the early chapters of this book. The story of Parzival is deeply connected with the questions raised by the activity of Jamshid and the extension of hands from the heart to cure the near-dead.

WAGNER'S VERSION

Richard Wagner, born with Venus in superior conjunction with the Sun, composed an opera titled *Parsifal*,[4] in which he pictured the dark magician Klingsor still in possession of the lance (the metamorphosed golden dagger) he had used to wound the Grail King. In Wagner's portrayal, the ailing King Amfortas as well as the Grail Knights depended upon the regular ceremony of the opening of the chamber of the Grail for their sustenance. In the interpretation of the San Francisco Opera in 2000, the knights were pictured as exhausted, mud-smeared soldiers in ragged battle outfits, indeed "the sick, the bleeding, the weak, the blind, the near-dead" from the Image at Scorpio. The events that began and ended World War II (chapter 5) come chillingly to mind. The knights were kept from death by the daily visitation to the Grail, yet barely so. No one faults the Grail—it is the source of life! But is it? Is this Jamshid's cauldron, and therefore ought we seek a greater king, not dependent on any technology whatsoever? Ought we depend on gestures from the heart through the hands? Indeed, in the tale of Parzival and in the opera of *Parsifal*, the deeds of the hero are required for liberation from the state of near-death.

The first performance of Wagner's opera *Parsifal* took place in the year of the last Venus eclipse, 1882, at 19 degrees and 2 minutes of Leo, thus square to the 20-degree mark of the Venus eclipse. Wagner's opera *Parsifal* opened on July 26, 1882. The Moon at the playing of the first note of music (8 p.m.) was at 20 Scorpio 23, exactly opposite the Venus eclipse of the Sun in 2012.[5]

---

4. Richard Wagner substituted the spelling Parsifal for Parzival, thinking that it came from the Arabic *fal parsi*, meaning pure fool. That etymology may have been false, as some scholars claim, yet it was inspired. The name Parzival can also mean "pierce the veil." One also finds the spelling Perceval.

5. Five hours later, at the end of the opera, the Moon lay at 23 Scorpio 11, thus mapping out the movement of the Venus eclipse between 2012 (20 Taurus 50, also the alignment activated on December 6, 1882) and 2004 (23 Taurus 5), linking Parzival and Wagner more intimately into our time.

Parsifal marks one of the most beautiful uses of opera—a word that means not one work, opus, but "the works," opus plural, every art available on display.

Tracing the lance in real life from the wounding of Christ Jesus to a museum in Austria, Trevor Ravenscroft popularized the tale of Parzival in his book *The Spear of Destiny*. He was born with Venus in superior conjunction with the Sun (noted at the end of chapter 5).

Parsifal emphasized the power of sexuality, Venus' gift manipulated for power—a realm with which Wagner had great familiarity, as we discussed in chapter 7 concerning *Tannhäuser*. Thomas Mann called Wagner's *Parsifal* a "sex opera of great daring."[6] In the performance by the San Francisco Opera in the year 2000, the second act opened with a large translucent scrim in front of the action, spanning the entire height and width of the stage, upon which was painted an immense skeletal pelvis. Everything was seen through the structure of an immense pelvis.

PRECURSORS

Rudolf Steiner named previous lives and reincarnations of individuals, tracing the development of the soul or individuality. Prior to Parzival, Steiner named the philosopher Mani, who was born with Venus in superior conjunction with the Sun (#5C, chapter 7).[7]

Prior to Mani, Steiner identified the youth of Nain, Martialis, who was resurrected from death, and thus born again, when the Sun lay at 23 degrees of Taurus, the site of the previous Venus eclipse of 2004. Thus Sun with

---

6. Quote from Raymond Furness, *Wagner and Literature* (Hampshire, UK: Palgrave Macmillan, 1982), p. 57.
7. See Richard Seddon, *Mani, His Life and Work, Transforming Evil* (London: Temple Lodge, 1996), chapter 9, which cites the unpublished lectures of Steiner, from which this series is taken. A new expression of Parzival is expected in our time.

Venus, and a relation to these degrees in Taurus (Scorpio), have a close connection to the story of Parzival and those who preceded him.

JAMSHID'S MAGIC TOOLS THROUGH TIME

We certainly see the powers of the lance, used for good or ill depending on the one holding the tool. The "spear of destiny" is ambiguous in its original use by the centurion at the foot of the cross. Some relate the tale as a demonstration of the callous attitude of the Roman soldiers. However, Anne Catherine Emmerich's version explains that the centurion on duty had actually become a follower of Jesus. When he learned that the officials were coming to break the bones of those being crucified, in order to hasten their death, he knew that this should not happen to Jesus' body. Though Jesus had apparently died, the centurion speared him, causing a large quantity of blood and water to fall onto the ground and into the centurion's eyes, curing a disease of his vision and opening his eyes to clairvoyant vision. Was Jamshid's dagger used for good or ill in this act?

Perhaps the most interesting feature of the Parzival story, brought forward to the Middle Ages from the time of Jamshid millennia ago, is the notion of keeping death at bay by the use of the Grail vessel.

KEEPING DEATH AT BAY

Amfortas suffered unceasingly. In a moment of weakness, he had been seduced by a beautiful woman and, in his vulnerability, speared in his side or thigh or testicles by the Black Magician, who used the lance that had pierced Christ's side on Golgotha.[8] As Gurnemanz related to Parsifal in Wagner's opera:

---

8. In Wolfram's version, speared in the testicles ("durch die heidruose," *Parzival*, 478–9, Book IX) in a joust, in Wagner's version speared through the side or thigh by the Black Magician. In classical thought, one can see the thigh as a generative organ in its own right, a point made very powerfully by Richard Orians in *The*

A wondrous lovely maid bewitched him.
In her arms he lay entranced;
The Spear dipped toward him—
A deathly cry!

. . . .

This wound it is that will never close.

The wound of the deathless dead leader has been unhealed and bleeding for years. There is a kind of imprisonment in the group of the Grail Knights as, understaffed, they try to carry out their role of protection of the Holy Grail, in a setting permeated by their leader's groans of pain. They await the one who can perform the deed necessary for the healing of the wound.[9] In the meantime they present the Grail in a daily procession that enlivens everyone. Contact with the wholeness heals, but just enough to get to the next day's revelation.

This gives a picture of life in Jamshid's world tinged with the pictures from the Image at 20–21 Scorpio—long life, yes, but not a pleasant one, just hanging on, suffering, "near-dead," yet not dying.

Though the body of the King lives, supported by the magical implement so similar to Jamshid's cauldron, generativity has diminished to a pitiful

---

Origins of European Thought (Cambridge, UK: Cambridge University Press, 1951), p. 182f. The use of *thigh* as an epithet for the genitals is also widespread: "The displacement of *genital* to *thigh* occurs frequently in the Grail myths" (Thomas McEvilley, *The Shape of Ancient Thought* [New York: Allworth Press, 2001, p. 669]). Also, in the Bible, as compare *yarek* in Gen. 24:2–9 and 47:29, translated as "thigh," to *yarek* in Genesis 46:26 and Exodus 1:5, named as the place from which offspring originate. Zeus gives birth to Dionysus from his "thigh." Carl Jung speaks of the "Amfortas Wound" in *Psychological Types* (par. 371–3), indicating the sexual nature of the location of the wound; cf. Marie Louise von Franz, C. G. *Jung: His Myth in Our Time* (Toronto: Inner City, 1998), p. 274. These references point to the common factor of Eros, that is, generative power.

9. Wagner imagines the Holy Lance that caused the wound can heal the wound by touching it. In just this way Telephus was healed by the lance that wounded him in the genitals—both the aggressor and the physician being Achilles.

state. Relations with women are distorted and nearly absent. The genitals are wounded and bleeding. The feminine has abjured this bunch.

When Parzival achieves the crowning as Grail King, Amfortas is released from his sufferings. One imagines that now he can die in peace.

We must wake up to the truth that death is not a disease. Death is a part of the development of one's soul via a learning sojourn on Earth. One comes into this world with an intention forged in starry worlds, and one works on that intention. When you've had a good try, you lay down your body with gratitude. It isn't a failure. From ignorance of the spiritual origins of our sojourns on Earth, from inadequate guidance about this, from distraction by the many delicious diversions, we develop the fear of death, and spend time, effort, and money in hopes of erasing any hints of it. But death is not a failure, nor a disease or something wrong.

Seen from the vantage of soul development, we can treat this whole ending differently. Let the experience of Amfortas inform us all—magical technology to keep us barely alive may not be in service of our spiritual pre-incarnational intentions for this life.

Some seek a premature death out of fear, to escape the pain of disease. This also disallows death's role in our development, a stage that we don't understand, just as adolescents cannot understand the real tasks of adulthood. It is not for us to second-guess how the stage of death benefits the development of our souls.

Modern Parsifal

Rudolf Steiner was twenty-one when *Parsifal* was first performed. He did not attend, but he heard about it, as it was a major cultural event. Soon after, Steiner became an editor for a magazine that commented on the artistic events of his time. Years later he spoke about the human soul as an archetypal plant, as if he were describing the brotherhood of the Grail: "In chastity and purity the plant stretches out its calyx toward the light, receiving its rays, receiving the Holy Love Lance, the 'kiss,' which ripens the fruit."[10] His teachings drew many people from all over the world to create and live together. He prompted the gathered Grail Community to build a Grail Castle, the Goetheanum, all working together in Switzerland while the artillery and bombs of World War I could be heard across the border.

In contrast, Adolf Hitler said, "I build my religion on *Parsifal*. Divine worship in solemn garb. Only in the robes of the hero can one serve God."[11] Hitler publicly imagined himself as Parsifal, even portraying himself in medieval armor on a large poster at Bayreuth, where *Parsifal* was being performed. However, Hitler secretly admired Klingsor, and created in the secret societies of Nazism a kind of Magic Castle based on black magic.[12]

The tale of Parzival or Parsifal has fascinated these two modern leaders of human beings, and many others besides, partly because the stories describe aspects of someone who really existed, who had multiple incarnations as a leader of humanity. Parzival, and later both Steiner and Hitler, gathered to themselves those of like-mind, those belonging to a particular karmic stream, which is the inspiration of Venus.

---

10. Rudolf Steiner, "Wagner and Mysticism," lecture of December 2, 1907, Nuremburg.
11. Quoted in Joachim Fest, *Hitler* (1973), cited in Raymond Furness, *Wagner and Literature* (Hampshire, UK: Palgrave Macmillan, 1982), p. 57. Hitler was born in 1889, well after the first performance of Parsifal.
12. See Jay Weidner, *The Mysteries of the Great Cross at Hendaye: Alchemy and the End of Time* (Rochester, VT: Destiny, 2003), for a compilation of the several forms which this took. Though Hitler sponsored extra performances of Parsifal early in his career, and made Wagner into the folk hero of the German state, after he rose to become leader of the country, he banned all performances of *Parsifal*, perhaps because the drama emphasizes comradeship with all, the power of compassion, and a distrust of Klingsor's magical powers.

9

SUMMARY: HANDS OF THE HEART

An event is occurring in the heavens, the only rare one in 2012. I don't deny that great changes are afoot in every aspect of life, from economic breakdown to violence in many large cities to plagues threatening and real. And I don't expect that June 5–6, 2012, will harbor a particularly powerful event. However, when looking at the issues at work in the themes related to the degrees where the Venus eclipse will take place, there is reason to become vigilant and respond to what could be a dramatic increase in technology that has the ostensible purpose of helping humanity—and the unintended consequence of enchaining it even more.

We are guided by the Image at Taurus and its opposite at Scorpio to develop capacities within rather than without, holding them in our heart. Indeed, we receive the hint that the forces of the will must emanate from the heart, with the light of one's interior streaming out in deeds to the world. We can respect the wonderful tools and magical implements that exist in the outer world, but learn to depend on the capacities of our own hearts. Reach out—not just metaphorically but physically—to other people and to the world. Reach out not to your computer, or to your pocketbook to give a donation to a worthy charity—but rather reach out to other human beings.

One feels perhaps the desire to perform big deeds for the world, as did Jamshid. However, the best way to give hands to the heart begins with small deeds. Here is an instance that I observed at an airport recently. The arriving passengers crowded around the baggage claim, spread out along two hundred feet of access. The bags were coming off the conveyor belt very quickly.

One bag fell off at a corner, at someone's feet. I watched her shift from a vigil for her own bags to a warm concern for an unknown passenger further down the line. She leaned over and picked up the fallen suitcase and set it back on the conveyor belt. Some people do this automatically. This woman had to shift her attention from her own concerns to become available to an awakening of her heart—the feeling for the plight of another. Then she could respond to what she felt. Her heart made the connection, and her hands did the work. She didn't require a magical tool, but simple hands from her heart. The example is intentionally small as these small deeds are accessible to everyone.

We have explored the ramifications of the Image at 20–21 degrees of Taurus, finding in the opposite sign of Scorpio clear hints about the heart. Venus' "eclipse" of the Sun in this degree will stimulate all of the themes that we've discussed, stimulate with a sense of the urgency of finding one's karmic group affiliations, those whose common purpose one has a destiny to encounter and further in the world. In Anthroposophy, such groups have sometimes been referred to as "streams," such as Grail Stream, King Stream, Shepherd Stream, Rosicrucian Stream, Templar Stream, and so on. One could add Gardener Stream, Inventor Stream, Mathematician Stream, Woodworker Stream, Social Worker Stream, and every other vocation to which humans are called.

In the context of this Image, one could feel the call of a powerful leader, such as Solomon, or Inanna, or Melchizedek. One could feel the call of one's Self. One could focalize around any of the technologies named, and find one's own way of accomplishing any of the miracles they appear to realize. One could define oneself by the group that shares a malady—bleeding, blind, near-dead. With what do you identify that is greater than yourself, that involves others now and through history? What is your soul group? Are you assisting in fulfilling its gift to humanity?

Your will-forces may well be activated by the Venus "eclipse" of June 5–6, 2012. Venus will excite these themes, and cast them in terms of karmic and soul connections. Each individual and the world as a whole will be challenged by great new technologies that will promise to release the world from disease, old age, and death. Will this lead to a narcissistic withdrawal each into his or her own heart, clothed in darkness? Or will each find the hands of the heart stretching forth in light, developing the tools of one's own being, the tools within the heart in a way that builds community?

From this Venus eclipse, new technologies will arise or be newly energized, and humanity will have to choose. This Image holds the potential for discovery that the faculties necessary for human development must be found in the human being, in the hands of the heart, not in technology. The cyborg—the human being embedded with machines—devolves the possible human being (the anthropos). Perhaps people will live longer but look more and more like those in the Scorpio Image, "near-dead." Remember that your life is meant to be lived for a purpose, and that death to this life is a natural stage in that development.

The distractions of convenience, technological gadgets that permeate our lives, must be set aside, at least for some period of time on a regular basis. If you cannot stop the intrusion of Jamshid's tools, then at least balance their effect with time in nature, in self-originated imaginations, in the exercise of the heart in the world of warm and real relationships.[1] One must have the

---

1. Two decades ago, Waldorf schools advised parents against any television for their children or any devices such as mobile phones until high school. Now they have retreated to advising against these things for the first seven, or now five, years of life. The teachers can point to studies, such as Jane Healy's *Endangered Minds* (New York: Simon & Schuster, 1999) or anything by Joseph Chilton Pearce, detailing the dangers of premature encounters with Jamshid's magic devices, but the pressure from the world eager for stimulation proves very powerful.

time to explore the vast realms of one's own heart, in a concerted practice of discovery and expression through simple deeds in the world. Find the capacities in your heart, then let the essence of your heart move out through your hands to the world and to others. Not conceptually, not even through miracle technological breakthroughs, but through your actual hands. Your attention is invited, as well as your participation in human relationships with others, beginning now, so you are not taken by surprise by the surges unleashed by the eclipse.

In order to achieve the most with this event, you can learn from the mentors who hover in this degree, and you can press aside the distracting influences of the spoilers who also hover in this degree (chapters 6 and 7).

Practice connecting heart to the light of the cosmos, as it expresses through the streams of destiny to which you are personally related—and to which you can come closer in this time. Let that light and power express through your hands. Thereby claim your place in space and time as an expression of that heart-filled light.

Revisiting the Magic Tools

Now we can revisit the magical tools which Jamshid wielded, and imagine how we could assume their powers through working the hands of the heart.

The golden dagger was used to master space by mapping it. You set out a diagram, as in a land survey, which becomes a concept of space in relation to the physical Earth. That means you can come back there again and, by finding the stakes or stones that you left, or by retracing the map you made, find the mapping of space again. We can learn to conceptualize the world with our own powers, to find our place there in relation to others working the land. We can develop a sense of Place through knowing the feeling of it. We can interact through the will, by touching the land and working its regenerative powers in gardening. This requires "high-sense perception," a more modern term than clairvoyance—clear seeing beyond the normal restraints of seeing, unbound by time and space—and clairaudience—clear hearing beyond the normal bounds of hearing, unbound by time and space—as well as the expansion of our other senses, into a clear knowing of the beauties of Sophia, the manifested beauty of life on Earth.

The magic cauldron helps us master time. We can see the past and the future through images that float in its waters. However, we can create these powers of high-sense perception on our own, without dependence on the machines that have become the repositories of our memory and our fantasy. To create this faculty requires practice in contemplation, meditation, intuition, all skills trained in different settings, including the Barbara Brennan School for Healing, Richard Bartlett's Matrix workshops, the School of Consciousness Studies at Rudolf Steiner College, as well as many other venues.

The firebird helps us change state, as in warming ice to water, and water to steam. It can refine conglomerations of things, so that metals of the same melting temperature are separated from the slag that remains. As we saw earlier, the firebird can be developed as our powers of attention, our ability to focus on something to warm it, or penetrate it, or understand it. This again is a form of the heightened powers that humans deserve to develop—and will not develop if overly dependent on technology to do so for them.

The magic ring helps us manipulate matter, in Jamshid's case by making it grow, and by extending our ideals into actual matter. The ring helps us generate not only pipe-dreams and pies-in-the-sky, but actual changes in the world. We can rely less on magical means to do this, such as advertising or political slogans, and turn to actual work in the world. Our two hands have remarkable powers to make things happen. When you join them with others of common will—the orientation to brotherhood and sisterhood stimulated by the sphere of Venus—it's astonishing what can be created. The hands must become active, following the advice and direction of the heart.

Emerson said it well: "Truly it demands something godlike in him who…has ventured to trust himself for a task-master. High be his heart, faithful his will, clear his sight, that he may in good earnest be doctrine, society, law, to himself."[2] Sovereignty offers the best possibility for growth as a human being, as well as the greatest responsibility to achieve it.

A Few More Recommended Activities

In addition to the many recommendations given above, here are some activities that we could recommend given that technology may receive a large boost from the Venus eclipse.

Assess the Kings and Queens in Your Life

Who's in charge of your life? To whom do you look up? To whom do you give your most precious asset, your attention? Observe yourself through the course of a day to recognize what claims your attention, both outer and inner. This may be more difficult than you assume. You will have to watch closely. Don't estimate at the end of the day. Rather set up a time, perhaps on the hour, where you observe what you're doing right now, and maybe note it in a little book.

Recognize what has power over you. Recognize the magical tools that surround you. Have the tools themselves become regal? Can you remind yourself of the true kings and queens—including the transitional ones, the Solomons and Inannas, as well as the king and queen of your own soul?

---

2. Ralph Waldo Emerson, *Emerson's Essays* (New York: Harper Colophon, 1926), p. 54, "On Self-Reliance."

You may say, "This technology helps me." Then I respond, Helps how? What is your life's goal and how is it helping achieve that? Have you become the servant of the technology? Or does it serve what you'd like to accomplish with your life?

Slow Degeneration and Rebuilding

Technology erodes human capacities, from the ability to imagine your own images vs. replay others' images, to the understanding of how to fashion wood and stone with the simplest tools rather than the complicated ones. Severely undercut is a capacity for clairvoyance, to sense beyond senses, or to expand sensory capacity into what Barbara Brennan called "high sense perception."

Take an occasional "fast from unlawful images"—that is, watch fewer movies. They give you unlawful images, because images of such great power ought to be something initiated from within you, not just brought up from the many movies that you have seen. When you take a fast, you know it's not forever, but you can balance the imbalance by having time with yourself occasionally, or in conversation with another—Venus listens, speaks, listens.

Erosion of capacity is something that elderly people recognize, and they don't like it. The losses imperceptibly accumulate, until you really can't do something that you could a few years previously. In a life cycle, this can be acceptable, as you shift your attentions and skills. However, too many young people are getting too old too soon. The many little losses to each individual and to our whole culture add up. These must be balanced by an easing of attention to technology as well as by a rebuilding process.

Rebuilding takes the form of daily practice. Even five minutes in the morning of sitting upright and clearing the mind of images assists. Meditative practices vary, and you can find the one that suits you. Five minutes in the evening is also helpful, to review the main features of the day, simply observing them, as you would observe a fish in an aquarium. By reviewing the events of the day, ideally backward, you digest them somewhat, and unravel the tightness of the day, and build your own capacities at noticing your own life. Do these building exercises daily, and they will counteract the degradations that we all suffer in the modern world. Spiritual beings are nourished by your contemplations, and in turn nourish you.

An exercise—pick up a tool. Take your relation to technology to its basics. Set yourself before a garden bed that needs weeding. Set out a few tools for the job. Weed by hand, then pick up a tool for weeding and use it for a few minutes. Set it down, and weed by hand for awhile. Note the subtle differences. Take this to other tools and other situations. There are many tasks that you can perform either with a tool or with your hands—picking up something heavy by hand, or with a hand truck, or with a bulldozer, for example.

Sit with a friend. Touch the friend's hand with yours. Touch his or her heart. Now do this with a tool of some kind, a stick or a spoon. How does that feel different?

Eat with your fingers. Alternate with eating with utensils. These are all ways of experiencing the subtleties of hands with or without tools.

Thoreau said that "men have become tools of their tools." Move it the other way, and reassert your mastery over your tools, and your ability to choose not to use them. Not always, sometimes. You choose.

Take Back Your Hands and Connect Them with Your Heart

As tools are an extension of the hands, so hands can become extensions of the tools. The balance has always tilted toward the former, where tools assisted the hands, which were for the most part involved in handling the stuff of nature, the wood and stone, and stalks of wheat, and the freshly ground flour, the dough for the bread, and the enjoyment in its eating. Now we have people whose hands know only a computer's buttons and a steering wheel. Their hands are becoming tools of the tools. Here a few suggestions to try out.

Practice handwriting. This seems like an odd recommendation, except that handwriting is no longer taught, rejected in favor of the printing of block letters because these can be scanned by computers. In a lovely little book, *Active Perception*, Alex Podolinsky presents samples of handwriting in German from the eighteenth and nineteenth centuries, and asks the reader to ponder these at length, demonstrating that each style of handwriting, even without knowing the content, reveals the soul of the writer. That's why computers cannot scan handwriting—something very intimate comes through this use of the hands.[3] Handwriting becomes an artistic activity. With pen in hand, write a letter to a friend, or a thank-you note, or a request to your mentor (in the case of the Venus eclipse, everyone shares the mentor of Mozart, chapter 6) to meet for tea. Several esoteric traditions recommend the technique of hand-writing a letter to a mentor or to the best aspect of yourself, then burning that letter as a means of sending the communication far and wide.

Touch Nature. Learn the textures of the barks of different trees, the feel of different minerals, the qualities of water by touch.

Garden with your hands—set a little plant in the soil and weed out the competitors around your plant. Cook with real foods that you harvest, wash, trim, slice, and set into a pot. Sometimes eat with your fingers.

Lift and carry stones. Build sandcastles at the beach. Inspired by Andy Goldsworthy's creations with natural substances, build things using your hands.

Touch people. Shake hands whenever possible. Learn to give others massage. Caress your loved ones—often.

---

3. Alex Podolinsky, *Active Perception* (Junction City, OR: Biodynamic Farming and Gardening Association, 2009). Rudolf Steiner also comments on this in *Approaching the Mystery of Golgotha* (Great Barrington, MA: SteinerBooks, 2010), p. 67, in a lecture from 1913.

Include your feet. Astrology credits the living being of the Fishes (Pisces) with the making of hands and feet. We have imprisoned our feet in shoes. You will find it amazing, when you go barefoot more often, to feel the different surfaces and what radiates from beneath those surfaces, even for five minutes a day. Steiner remarked that two degenerations marked the coming of modern times: the degeneration of handwriting (explained above) and the invention of the motor car.[4] In another place, cited in chapter 7, he said that he was happy with the motorcar and used it often, adding that one must balance one's time in a motorcar with time out of it. In other words, walk from place to place, at least some of the time. In Nature, one feels better about taking shoes off, as what the feet feel seems healthier. Thus there are reasons to cover your feet with shoes, and also reasons to have them off again. Wiggle your toes, grasp the earth with your toes, awaken your feet.

Touch other people's feet! Here's a combination of suggestions. You learn foot massage or reflexology, and then can offer to massage other people's feet, thus awakening your hands and their feet at the same time.

Build a Temple of the Heart. This may come naturally, or you may have to train yourself in some of the ways suggested above, so you are ready when the opportunity arises, as it will often come without warning—an opportunity to assist another. Hands-from-the-heart has a prerequisite: a sensitive heart, a heart that listens, thus accentuated by the presence of Venus. We can learn from the Images from both the Bull and the Scorpion/Eagle that the opportunities many increase in frequency. You can train this faculty at any time. Cultivate feeling. You do this by listening to others, listening with warm interest. That formula can be your rule of thumb—"warm interest." Cultivate devotion or reverence, best done in relation to some other living

---

4. One of many prescient comments by Steiner, this one from *The Evolution of the World*, op. cit. Who would have predicted that handwriting would no longer be taught in schools? Yet that is what has happened.

thing—flowers, animals, other human beings. You may note how much time you spend devoted to your machines. Balance this with attention to amazement and wonder at life. Devotion to life builds a temple which Venus can easily enter. Develop not the cool concept of devotion, but the passionate cultivation of active warm love. Venus may enthrall us with Jamshid's magic tools, and direct our attentions to their many wonders. We fall victim to the erotic phantasm of the latest gadget. Do not reject her for this seeming betrayal, for Venus/Eros also gives us the strength to empower our own beings with love, and then reach out to another.

Hands from the Heart. Go to a place where there is need. This doesn't mean a place where gather the needy, as in certain parts of any city where there are many hands thrust toward you asking for money. That's graduate school for the one who develops hands-from-the-heart. Begin with a place where there is need, but not such pressure on every passerby. Many settings will do. Many nursing homes give opportunities for arousing your feelings and then connecting that with your hands.

Music Helps. In this study, Mozart and Wagner front for music in general, ideally music played by present hands on present instruments and not on a recording. Sing. Everyone has a voice that can sing, no matter what you've been told in school. Sing. You will discover how sound moves out as invisible hands coming from your heart. Improvisational singing expresses the will in the most beautiful way.

Listen—sing—listen.

APPENDICES

APPENDIX A: CELESTIAL PHENOMENA

Alignment of Sun, Earth, and Galactic Center

For a complete explanation of the dynamics of Sun, Earth, and Galactic Center, I have written a paper that has been published in a few places.[1] Here I will give a summary.

An alignment of Sun and Earth to a point *near* the Galactic Center happens twice every year. That looks like this: Sun—Earth—(near) Galactic Center, or like this, Earth—Sun—(near) Galactic Center.

As our solar system lies over ten light-years above the main disc of the galaxy, and the Sun–Earth plane (the ecliptic) is tilted at 60 degrees from that plane of the galaxy, technically the line between Sun and Earth does not point to the Galactic Center at any time. It will point either to the place on the great circle perpendicular to the ecliptic which is closest to the Galactic Center (2 degrees Sagittarius) or to the place where the Galactic Equator, the plane of the Milky Way, crosses that of the ecliptic (5 degrees Sagittarius).[2]

You can look at the calendar date when this alignment occurs. Owing to the slow precession of the equinoxes, the alignment occurs at different times

---

1. "The Galactic Center and December 2012," in *Journal for Star Wisdom 2010* and in the *Journal of the International Society for Astrological Research* in 2011, and now at the www.StarWisdom.org website.
2. These degrees are from sidereal reckoning, which is their actual astronomical location along the ecliptic (Sun–Earth plane) in a system of twelve equally long zodiacal signs. However, the point made here does not require that one embrace a sidereal view of the heavens (vs. the conventional tropical or seasonal view).

in relationship to the seasons. At this time the alignment occurs at the Earth's solstices (December 21, shortest winter day in the northern hemisphere, longest summer day in the southern hemisphere, and June 21, just the opposite day lengths). As the Sun is half a degree across, the alignment of Sun, Earth, and (near to) the Galactic Center occurs twice a year for a period of over thirty years before the Sun—Earth—(near) Galactic Center alignment moves slowly to the twentieth of December and June, then the nineteenth, and so on. From the point of view of the Galactic Center, there is nothing rare about the alignment on the days of the winter or summer solstices.

The prediction of the movie *2012: Doomsday* that there will be five celestial objects in a unique alignment, causing the end of civilization and the death of all but a few thousand people, will not be met in the year 2012.

Prophecies Pointing to 2011 and 2012

I will not reference the dozens (and perhaps hundreds) of books and videos that predict the end of the world on December 21, 2012. However, I do note that some researchers trace the end of the Mayan calendar to October of 2011. Twenty years ago Ken Carey predicted great changes in these years, indeed, a complete shift from fear-based emotions to a world founded on love between everyone and everything.[3] Writing in the middle of the twentieth century, Alice Bailey also mentioned this year.

Some have pointed out that the planets Uranus and Pluto will be square to each other, which is also rare. However, the encounter is spread out over time. It will occur seven times from June 24, 2012, to March 17, 2015. Using the fifteen-degree orb that Richard Tarnas prefers for outer planet aspects, the square between Uranus and Pluto will extend from the middle of 2004 to the middle of 2020.[4]

3.   Ken Carey, *The Starseed Transmissions* (New York: HarperOne, 1991).
4.   Richard Tarnas, *Cosmos and Psyche: Intimations of a New World View* (New York: Plume, 2007).

In contrast the Venus eclipse of the Sun will be within a fifteen-degree orb from May 26, 2012, to June 15, 2012, though its effects will extend further in time before and after the event.

Theories based on the end of one of the several Mayan calendars in use from that civilization emphasize the end of a historical period spanning thousands of years. In comparison, the Venus eclipse of the Sun is less rare, but has the virtue of being an actual astronomical event.

ANNULAR ECLIPSES OF THE SUN

An annular eclipse by the Moon occurs when the Moon is further away from the Earth and does not completely cover the face of the Sun. One observes a fiery ring of Sun around the center darkened by the Moon. Thus the Venus eclipse is a version of an annular eclipse, though the rim of the Sun around the dark dot of Venus is huge (as Venus is 1/112th of the Sun's diameter). The notion of a seven-ringed cup can also stimulate this imagination: that there are seven rings of increasing size, indicating not the bodies of the seven classical planets, but the rings of their orbits.

Rudolf Steiner referred to the Venus "eclipse," giving hints but little detail: "Very significant things can be observed when Venus is passing in front of the Sun. One can see what the Sun's halo looks like when Venus is passing in front of the Sun. ["Halo" is another way of speaking of an annular eclipse of the Sun by the Moon.] This event brings about great changes."[5] He did not say what those changes might be, though he suggested that the event would influence the weather.

---

5. Rudolf Steiner, *From Sunspots to Strawberries... Answers to Questions* (London: Rudolf Steiner Press, 2002), pp. 170–173.

Transits of Venus

The height of the Venus eclipse will be around 1:00 a.m. UT on June 6, 2012, which is 11:00 EST Australia on June 6 (and late on June 5 in the United States). The duration will be 6 hours and 40 minutes.

The time between the pair of transits, what I call eclipses in this book, is eight years or 2^3 years. The time between the first of the previous pair (as in December 9, 1874, both at 23 degrees and 58 minutes of Scorpio) and the first of the present pair (to June 8, 2004, Sun and Venus at 23 degrees and 2 minutes of Taurus) is 129 ½ years = $5^3 + 5 - .5$.[6]

Looking at longer stretches of time, the gap of 129.5 years oscillates with a gap of 113.5 years, totaling 243 years, which repeats. $113.5 = 5^3 - 5^2 + (3 \times 5) - (3 \times .5)$. Thus the time between the last of a pair and the first of the next pair alternates between 121.5 and 105.5 years. $121.5 = 5^3 - 5 + (3 \times .5)$ and $105.5 = 5^3 - 5^2 + 5 + .5$.[7]

The sidereal orbit of Venus around the Sun is 225 days, which is $5^3 + (4 \times 5^2)$, also seen as $5^2 \times 3^2$. Seen both from the points of view of the different cycles of year and day, the number 5—and the powers of 5, its square and cube, and fraction—plays an important role in the dance of Venus with the Sun. In any eight-year period (2^3), Venus completes a five-petalled flower from the Earth's point of view, having in that time moved on a small increment through the zodiac. The five-petalled flower is particularly beautiful, and is inscribed in stone at the dance ground immediately to the South of the StarHouse.[8]

---

6. The same can be said from the second of the previous pair, December 6, 1882 (with Sun and Venus at 21 degrees and 26 minutes Scorpio), and the second of the present pair, June 6, 2012 (Sun and Venus at 20 degrees and 51 minutes of Taurus).
7. Looking over many centuries, one occasionally sees that the second of a pair does not occur, indicating another, longer cycle in the Venus orbit of the Sun.
8. From Joachim Schultz, *Movement and Rhythms of the Stars* (Edinburgh: Floris, 1963), p. 121.

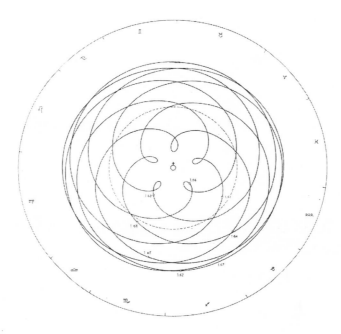

In contrast, transits of Mercury occur much more frequently. November transits occur at intervals of 7, 13, or 33 years; May transits only occur at intervals of 13 or 33 years. Combined, there might be only three years between a May and a November transit.

If you connect the five inner petals of the figure above, you form a pentagon. If you connect every other one, you make a five-pointed star. These have interior angles of 72 degrees (5 x 72 = 360, the full circle). From the Earth's point of view, Venus follows the Sun, then disappears into an inferior conjunction, then leads the Sun to the west. The time between these two maxima is 144 days, twice 72.[9] Thus 72 relates to degrees of separation in space, as well as to time.

---

9.   Noted in Robert Graves, *White Goddess*, op. cit., p. 251.

APPENDIX B: JAMSHID IN GREATER DETAIL

Jamshid has been called Jamshed, Jamshyd, Jambushad, Janbushad,[10] Djem, Djemschid, Dschemschid,[11] Djemscheed, Gamshed, Dsemchid, Djemjid,[12] and Djemjdid.[13] In the Avestan language of most ancient Persia, this is the shining aspect (*-shid*) of Yima. One can sometimes find Yama used instead of Yima. I use Jamshid for all the different spellings of this ancient name.

One of Jamshid's names, Yama, recalls the Sanskrit word Yama, meaning qualities of character that come from restraint of the bestial temptations of the lower nature of the will. It has five qualities:

1. Ahimsa, nonviolence, choosing not to commit the violence to which one is drawn,
2. Satya, truthfulness, choosing not to deceive despite the urge to do so,
3. Asteya, non-theft, choosing not to steal from others,
4. Brahmacharya, moderation in sex, and sometimes abstinence, choosing not to rush into sexual pleasures, and
5. Aparigraha, non-acquisitiveness, choosing to give away wealth and to ignore desires.

These qualities of Yama, which may have been held over from the Indian epoch to the Iranian epoch when Jamshid flourished, may give an idea of the character of this great king, a mastery of the will—which we have said technology expresses.

---

10. These two from Giorgio De Santillana and Hertha von Dechend, *Hamlet's Mill*, op. cit.
11. From Walter J. Stein's *The Ninth Century and The Holy Grail*, op. cit.
12. Rudolf Steiner, *According to Matthew*.
13. From the explanation of the title of the annual series *The Golden Blade* (Edinburgh: Floris Books, periodical).

The name Jamshid calls forth the participation of the celestial creators in the "fixed signs" of the zodiac, Taurus and Scorpio, which we have seen, as well as Leo and Aquarius. Working backward in that name, "d" is the gift of Leo, "sh" the gift of Scorpio, and "m" the gift of Aquarius. If one pronounces "j" as in "jump," the sound "j" comes from a combination of "d" (from Leo) and "zh" (from Scorpio). Only the gift of Taurus, "r," is missing in this name. The vowels of "ee," mediated into our world by Mercury, and "ah," mediated by Venus, become the bearers of the consonants of Leo, Scorpio, and Aquarius into the human realm.[14]

One can also feel a similarity in sound between the core name, Yama or Yima, and the declaration, "I am!"

Rudolf Steiner affirmed that the gifts to Jamshid came directly from Ahura Mazda, the Father God, who taught Zarathustra five thousand years before the battle of Troy. Steiner referred to Troy because that confrontation between the Greeks from the west and the oriental peoples from the east broke the last vestiges of the ancient Persian culture, making way for the epoch of Greece and Rome.[15] Zarathustra had laid the foundations for the Persian culture in its beginnings, and then, in a later incarnation as Zoroaster (the same name in a different language, also meaning golden— *zoro*—star—*aster*) in Babylon, instructed the Jews who had been brought there in the esoteric foundations that they then turned into mystical Judaism. He also founded the modern school of astrosophy with Aldebaran–Antares

---

14. With the spelling of Jamshid as Yima, the only consonant, "m," comes from Aquarius, and the rest are the vowels of Mercury and Venus. The association of sounds with gifts from the zodiac and planets originated with the Hindu understanding of the matrika goddesses and was picked up and developed by Rudolf Steiner in the art of eurythmy.

15. The history, and gap of five thousand years, came from Plutarch's *Isis and Osiris*, 46–7, which Steiner affirmed by his use of this evaluation of time. Between the time of ancient Persia and the Greek/Roman times came the 2,160 years of the Egyptian/Chaldean civilization.

as the main axis dividing the heavens, giving 1 degree Aries—exactly 45 degrees, or half of 90 degrees—as the starting point.

Steiner spoke of Jamshid's leadership of the whole Iranian or Persian epoch, the second post-Atlantean civilization, after ancient India and before the age of Egypt/Chaldea. Through Jamshid's guidance, the Iranians did not follow the path of their neighbors, the Turanians, who tried to use the decrepit magic of Atlantis to change substance and influence people. Their old spells turned dark, and, though they wreaked havoc on their neighbors in wars supported by black magic, the misuse of these powers folded in upon itself, and the Turanians eventually died out.[16] You can see a picture of these two realms in conflict with each other in Tolkien's trilogy *The Lord of the Ring*, and indeed one could see Jamshid's ring pictured there, too. One can see the contrasts of these ancient tales in the contrast between the pictures of the Scorpio and Taurus Images.

In Jamshid we see a figure similar to Marduk of the Sumerians and Prometheus of the Greeks, a bearer of fire, teaching how to make it and how to use it for many crafts including metalworking, the fashioning of glass, the arts of agriculture, and intelligence generally. He has the ability to apply intelligence to the earth, measuring it, dividing it, and cultivating it. The golden dagger becomes that which separates, categorizes, measures, stirs and plows the earth. It is the prototype of all tools and inventions of civilization. Separation is the task of the senses. From the blob of color and sound that the infant experiences, we slowly differentiate between one kind of stimulus and another as we develop our intelligence, exercising the knife of discernment.

A person with a connection to this Image might speak aloud some of the words of Jamshid himself, such as those quoted above, where Jamshid speaks to the Creator, Ahura Mazda, "Yes! I will make thy worlds thrive, I will make thy worlds increase. Yes! I will nourish, and rule, and watch over

---

16. Steiner, *According to Matthew*, lecture 1.

thy world. There shall be, while I am king, neither cold wind nor hot wind, neither disease nor death." Speak it loudly, as a proclamation. Speak it softly and tenderly, for Ahura Mazda and the Seven Rishis are also close by, and can hear your words of intent when spoken quietly.

APPENDIX C: OBSERVATIONS
ON ASTROLOGY AND ASTROSOPHY

I would like to share some perspectives on astrology and its more mature cousin, astrosophy. I cannot give a complete presentation of the field in this space, nor give a full credo or defense, yet I would like to answer the repeated demand, "How can astrology possibly work?! The planets and stars are so distant! Show me the mechanics of this supposed effect." I have heard some astrologers answer this challenge, "Well, it just works, that's all." I can add some observations to that unsatisfactory answer.

A MECHANICAL POINT OF VIEW

A materialist has the perfectly practical and understandable desire to understand the workings of something, to have it explained and demonstrated, a point of view that is often summarized "seeing is believing," or "what you see is what you get." This serves well the getting from one place to another, and handling many of the requirements of life. Without rejecting this mode of thinking, which is very useful in many settings, one can ask the materialist how iron filings arrange themselves on a paper when a magnet is passed beneath the paper. One can ask how it is that a dog knows when its owner is returning home, minutes before he or she comes in the door. One can ask how clairvoyants have seen the locations of lost children and hidden treasures. One can ask a dozen different questions, not to reject the materialist point of view, but simply to demonstrate that it is one point of view among many. And it has not served its adherents well, who two hundred years ago would have rejected as impossible the powers of X-rays, gamma rays, and homeopathy. The challengers are not scientists, who are ideally and often open-minded, but rather those who insist on seeing how things work.

A recent study that I found most interesting in this regard involved a firewalk in Spain, a regular and very sacred ceremony of a small community there. A scientist was able to get permission to put heart monitors on those walking over hot coals, and on several members of the audience, seated ten yards or more from the fire. Well before the event, there was no correlation between the two. During the event, the heart rate of the one walking over the red-hot coals correlated very highly with his or her friends and relatives, but not with unrelated spectators. We resonantly connect with those to whom we are related by blood or attraction.[17]

Sun and Moon

The Sun is distant, yet many awaken to the beginning glow of daybreak without the use of an alarm clock. At different times of the day, we feel differently and think differently. The proverb "Mornings are wiser than evenings" has to do with the cycles of the Sun. Yet there are no strings running from the Sun to our behavior. Likewise for the Moon, whose phases determine the cycles of all living things.[18]

---

17. "Synchronized arousal between performers and related spectators in a firewalking ritual," *Proceedings of the National Academy of Sciences*, Ivana Konvalinka et al., 2011; http://www.pnas.org/content/early/2011/04/26/1016955108.full.pdf+html.
18. *Moon Rhythms in Nature*, by Klaus-Peter Endres and Walter Schad (Christian von Arnim, Trans., Edinburgh: Floris, 2002), gives dozens of examples from the scientific literature.

MEDICAL MODEL

Materialists like to point to medical research as a model that astrology should copy. Yet that model is seriously flawed. Pills without anything in them—called placebos—have been shown to be as effective, and in some cases, more effective than the pills with drugs in them. As only the studies showing a positive effect for the drug are published, the picture is illusory.[19] Medical research is often flawed by manipulated data.[20]

Agriculture is also seen within this scientific medical model. Add something to the soil, and it has thus-and-such an effect, because of a defined "mode of action," such as added magnesium making the manufacture of chlorophyll easier. Then there are such practices as biodynamics or the use of ground up rock powders whose positive effects cannot be explained. In a session on rock powders at a conference on agricultural techniques, a senior professor at the Department of Agriculture of Kassel remarked to me, "We don't know why finely ground rock powders work—we think it's sheer energy, which is no explanation at all!" Lack of explanation for mode of action is more common than rare, in all fields into which science inquires, making astrology less unusual.

When you read the history of most drugs, you find a kind of experimental approach—let's try this, let's try that. A chemical has many effects in a system, be it agricultural or your body. The medical industry latches onto one of those effects, as if it were intended all along, and calls the others "side-effects." At your next visit to the doctor's office, read through the Physician's Desk Reference on any drug with which you are familiar—even salicylic acid or aspirin is there. You will see a list of side-effects that is in tiny print, partly because the manufacturers don't wish you to know about all these

---

19. *The Emperor's New Drugs: Exploding the Antidepressant Myth*, by Irving Kirsch (New York: Basic, 2011).
20. As, for example, reported in "An Array of Errors," *The Economist*, Sept. 10, 2011, http://www.economist.com/node/21528593.

side-effects, and partly because, if printed in normal type, it would overwhelm the content of the intended effects.

The ideal of the medical model is a pillar of rational materialistic thought, and it often fails. Let's contemplate why.

Causation

Perhaps the problem is with an understanding of causation. Something is seen as a cause and something as an effect. As in the discussion of the medical model, we see that every cause goes in many different directions, but we like to see one of them land on effect, which can also have many causes. It's our way of understanding the world.

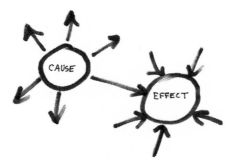

In this model of causality, we would say, "Venus made me do it." But that undermines the inherent freedom of the soul, and the facts of the matter. More accurate would be the notion of a *matrix of intercorrelated phenomena*, each phenomenon or object (A through E in the diagram on the next page) complicatedly interwoven, each a cause and each an effect of the others.

Our task is to enter the system and describe its objects and its relations as best as possible. This requires an honest approach, one where each must hold illusion at bay.

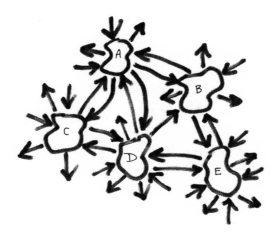

The new experiments in physics demonstrating entanglement illustrate this dilemma. Two particles linked together are split, and then observed two inches from each other, which on the scale of these experiments is a long way. It could be two miles from each other, to speak to our human sense of distance. When one particle changes its state or phase or orientation, that is, behaves differently, it has been observed that its twin does exactly the same at the same time. Einstein was aware of entanglement as a prediction, and didn't like it, calling it "spooky action at a distance." Entanglement has now been demonstrated with many kinds of particles. Where are the strings sought by the materialist?

The theorists in this field suggest that everything is connected with everything else, that indeed we live, move, and have our being in a matrix of intercorrelated phenomena. We can observe trends or tendencies, but no amount of distance, or scale differences, as in small versus large—the main arguments against astrology—is a boundary against relationship.

A Mental Approach to Zodiacal Signs

Professional astronomers wanted to divide up the sky into sections so that they could give celestial addresses to what they saw in their telescopes. Several systems have been devised over a few thousand years, and I give (opposite) the map most recently devised by the international organization of astronomers. They did not connect dots, because they wanted the stars inside of a zone or place, not on its edges. They went around the stars as well as they could, using right angles to make location of stars easier for them. They used some traditional names of the zodiac, but inserted other ideas. The chart gives the calendar months in the top line. Next appear the astrological signs using the tropical method (showing their lack of correlation with the actual constellations). The property map shows the way that the astronomers have carved up the sky. Not shown are the equal-length signs of Zarathustra (astrosophy), which, however, can be easily imagined as relating to the actual constellations. The point of including this map is to show that the astronomers who made it do not appear to have been sensitive to the living beings of the heavens. It looks more like a division of voting districts by the political party in power so that they more easily win the next election.[21]

These are some answers to questions that I have received numerous times. Though not a complete defense or apologia, they may make astrology and its more mature counterpart, astrosophy, more understandable. I would answer the materialist's formula, "What you see is what you get," with the

---

21. Illustration from the *Cycles* newsletter of The School for Ageless Wisdom in Arlington, Texas, http://www.unol.org/saw2/cycles/index.html. The technique is called gerrymandering, meaning making something look like a salamander, after the first politician (Eldridge Gerry) to do this.

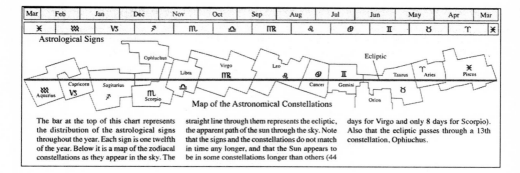

Mar	Feb	Jan	Dec	Nov	Oct	Sep	Aug	Jul	Jun	May	Apr	Mar
♓	♒	♑	♐	♏	♎	♍	♌	♋	♊	♉	♈	♓

Astrological Signs

Map of the Astronomical Constellations

The bar at the top of this chart represents the distribution of the astrological signs throughout the year. Each sign is one twelfth of the year. Below it is a map of the zodiacal constellations as they appear in the sky. The straight line through them represents the ecliptic, the apparent path of the sun through the sky. Note that the signs and the constellations do not match in time any longer, and that the Sun appears to be in some constellations longer than others (44 days for Virgo and only 8 days for Scorpio). Also that the ecliptic passes through a 13th constellation, Ophiuchus.

response, "Then learn to see better." Expand your vision to include what can be felt, that is, in the terms of this book, learn to see with the heart.

APPENDIX D:
STAR CHART FOR THE VENUS ECLIPSE 2012

This star chart for the Venus eclipse of 2012 shows the starry background of the movement of Sun and Venus and displays Zarathustra's vision of the organization of the cosmos. The symbols in this chart are familiar to many, though some readers may not know them. You can find the symbols of the zodiac named in the chart at the end of Appendix C. The Sun is a circle with a dot in the center. Venus is a circle with a cross beneath. The number 2 shows the location of Antares, the Heart of the Scorpion. Aldebaran lies directly opposite, at the center of the Bull, Taurus. The number 5 shows the location of Spica, the Goddess Star, conjunct with Saturn at the end of the sign of the Virgin (Virgo). The number 9 at the top of the chart (which may appear as a 6) shows the location of the Galactic Center discussed briefly in Appendix A.

Find out much more about this approach from www.StarWisdom.org.

Hist Evnt of Venus Transit 2012 - Geocentric
At Greenwich, Kent, United Kingdom, Latitude 51N29', Longitude 0W0'
Date: Wednesday, 6/JUN/2012, Gregorian
Time: 0:57, Time Zone GMT
Sidereal Time 17:56:23, Vernal Point 5✕ 5'12", House System: Equal
Zodiac: Sidereal SVP, Aspect set: Conj/Sq/Opp WIDE

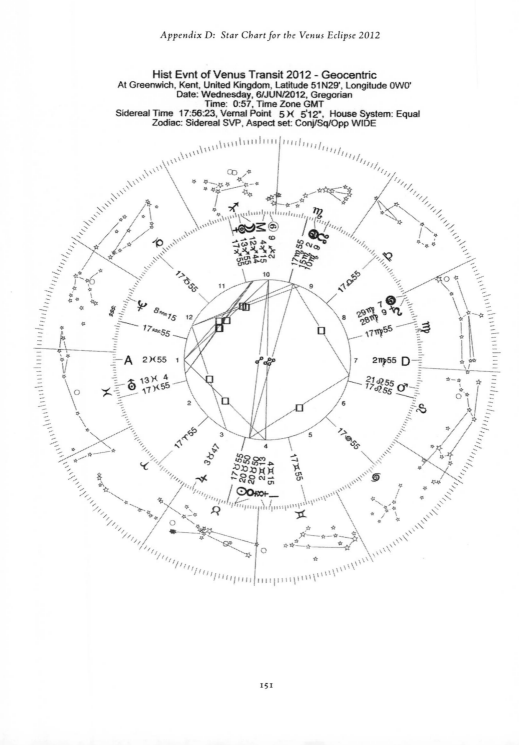

STAR WISDOM AND RUDOLF STEINER
A Life Seen through the Oracle of the Solar Cross

David Tresemer, Ph.D., with Robert Schiappacasse

A gift was given you at your first breath—have you opened it? It came from the stars and impressed itself into you at your most vulnerable moment. In this introduction to astrosophy, or star wisdom, a science with ancient roots and modern relevance, David Tresemer shows how the patterns written in the heavens influence a person's life. Taking as an example the remarkable life of Rudolf Steiner, Tresemer demonstrates the Oracle of the Solar Cross, whose four points interact throughout each of our lives to shape opportunities and challenges that our souls must face.

Rudolf Steiner spoke often of the "starry script" and hinted that, whereas its patterns impress themselves upon the human psyche, we can also influence this heavenly text, altering it in service of the continued development of humanity. Through stories from the life and examples from the work of Steiner as well as the "Star Brothers and Sisters" who share aspects of his Solar Cross, Tresemer illuminates this truth.

Rich in art and anecdote, this groundbreaking book gives insight into the foundations of Anthroposophy and shows how great acts, feelings, and thoughts by human beings on Earth shine out and impress their patterns into the cosmos.

Robert Schiappacasse has been a student of Rudolf Steiner's Anthroposophy for more than thirty years. He developed a deep interest in humanity's relationship to the world of the stars and, in 1977, began studies with Willi Sucher, a pioneer researcher in the field of Astrosophy, or star wisdom. He presents at conferences and workshops on star wisdom themes and other anthroposophic topics. He is coauthor with David Tresemer and William Bento of the book *Signs in the Heavens: A Message for our Time* and has coauthored articles with David Tresemer, including "The Chain Reaction Experiment"; "The Signature of Saturn in Christ Jesus' Life"; and "The Signature of Pluto in the Events of Christ Jesus' Life."

www.steinerbooks.org | ISBN: 9780880105743 | 396 pages | paperback | $25.00

ONE TWO ONE
A Guidebook for Conscious Partnerships, Weddings, and Rededication Ceremonies

Lila Sophia Tresemer and David Tresemer

"David and Lila invite us to explore and express our deepest yearning to be fully and vibrantly who we are in our sacred relationship. In service of this they bring tools—both concrete and mystical—through which they mentor us each step of the way."
—**Shana Parker**, Ph.D., Heart Vision/Licensed Psychologist

The founders of The StarHouse in Colorado bring us this guidebook for all aspects of intimate partnerships—beginnings, recommitment, and even healthy endings. A relationship is an opportunity for growth an spiritual maturity, and the authors provide exercises for partners to explore themselves and each other more fully within the context of intimacy. *One Two ONE* includes innovative tools for designing rituals (weddings or other) that best express individual and universal aspects of loving relationships.

Lila Sophia Tresemer is a group facilitator, playwright, photographer, writer, ceremonialist, and minister. Lila cofounded The Path of the Ceremonial Arts program at the StarHouse in Boulder, Colorado, now in its tenth year, supporting women to remember, heal, and transform through the ritual arts. She has produced and directed *Brain Illumination*, an animated DVD journey for personal and planetary enlightenment, and produced and coauthored *Couples Illumination*, an animated DVD dedicated to creating conscious partnership (see next page). Lila cowrote and produced a trilogy of plays on the Divine Feminine, including *My Magdalene* and *Darwin in the Dreaming*. She also cofounded the Women of Vision outreach to the women of the Middle East in 2004, which continues to bring together women from many traditions (Arabic, Jewish, Muslim, Bedouin, Druze, Christian, earth-based traditions), both at the StarHouse and in Israel/Palestine.

www.steinerbooks.org | ISBN: 9781590561652 | 268 pages | paperback | $24.95

COUPLE'S ILLUMINATION
Creating a Conscious Partnership:
An Animated, Energetic Look at
How Relationships Can be More Successful DVD

LILA SOPHIA TRESEMER AND DAVID TRESEMER

True partnerships between people are an essential element of a joyful existence, yet so rarely do we have examples and practices for creating them consciously, with joy and ease! We often spend more time arguing and negotiating relationship— rather than loving, laughing and growing!

Learning about Conscious Partnership can orient you to each other in a new way, and can acquaint you with the larger fields of energy and Love from which you both originated. Through the many practical tools given here, your relationships can grow. You can learn how to see more clearly, act with integrity, and communicate with respect, thus healing separation between you and your partner. Conscious Partnership is a pathway to the matrix that connects you with the wonders of life all around.

Part 1: Offers wise words from Lila and David, illustrated by animations and accompanied by specific exercises.

Part 2: Provides an animated meditation which will support a continued practice in developing your relationship.

This DVD acts as a companion to the book, *One-Two-ONE: A Guidebook for Conscious Partnerships, Weddings, and Rededication Ceremonies*, presenting some things more effectively than a book can do.

Vist www.davidandlilatresemer.com for more on books, events, and courses from David and Lila Sophia Tresemer.

www.steinerbooks.org | ISBN: 9781590561706 | DVD | $24.95

David Tresemer, Ph.D., has a doctorate in psychology. In 1990, he cofounded the StarHouse in Boulder, Colorado, for community gatherings and workshops (www.TheStarHouse.org) and cofounded, with his wife Lila, the Healing Dreams Retreat Centre in Australia (www.healingdreams.com.au). He also founded the Star Wisdom website (www.StarWisdom.org), which offers readings from the Oracle of the Solar Crosses—an oracle relating to the heavenly imprint received on one's day of birth. Dr. Tresemer has written in many areas, including *The Scythe Book: Mowing Hay, Cutting Weeds, and Harvesting Small Grains with Hand Tools* and a book on mythic theater, *War in Heaven: Accessing Myth Through Drama*. With his wife, he coauthored the recent *One Two ONE: A Guidebook to Conscious Partnerships, Weddings, and Rededication Ceremonies*. With William Bento and Robert Schiappacasse, he wrote *Signs in the Heavens: A Message for Our Time*, about the comets Hyakutake and Hale-Bopp and their crossing of the mysterious and ominous star Algol. David also coauthored with Robert Schiappacasse *Star Wisdom and Rudolf Steiner: A life Seen through the Oracle of the Solar Cross*.